BEHAVIOUR OF MARKET PRICES OF AGRICULTURAL COMMODITIES

BEHAVIOUR OF MARKET PRICES OF AGRICULTURAL COMMODITIES

S.R. Takle
Principal and Head
Department of Economics
Narayanrao Waghmure College Ak. Balapur, Maharashtra

V.B. Bhise
Professor, Department of Economics
Dr. Babasaheb Ambedkar Marathwada University
Aurangabad

SERIALS PUBLICATIONS
New Delhi (India)

First Published - 2007

ISBN: 81-8387-074-0

Published by

SERIALS PUBLICATIONS
4830/24, Prahlad Street, Ansari Road
Darya Ganj, New Delhi-110002 (India)
Phone : +91-11-2324 5225. Fax : +91-11-2327 2135
E-mail: serials@satyam.net.in

CONTENTS

PREFACE

Agricultural prices play an important role in national economy of India. Prices of agricultural products particularly the foodgrain and oilseed prices play an important role not only in the agricultural sector but in the whole national economy of India. Agricultural prices do affect agricultural production decision by the farmers, intersectoral terms of trade, cost structure and income of the farmers.

Agricultural prices is a big problem in Indian agricultural sector as agricultural production depends upon monsoons. Agricultural prices are subject to fluctuations any time. Agricultural prices variation is an important issue in Indian agriculture. Agricultural price vary from places to place at a point in time as well as from time to time.Such price variations are regarded as spatial and temporal price variations respectively.

The present study is an attempt to measure and analyses the temporal variation in prices of foodgrains and oilseeds at regulated markets. It provides a brief review of study on price variation of foodgrains and oilseeds conducted in different part of India. The main focus of the study is on analysis of seasonal, annual and long - term variations in prices of agricultural commodities at regulated markets. Further, it also studies the relationship between prices and arrivals of agricultural commodities at regulated markets. An attempt has been made to make suggestions to minimize the variations in prices of agricultural commodities in India. The statistical analyses made in the present study broadly covers the period1981-1995, and is based on secondary data drawn from office records of the regulated markets in the study area.

This work has bean completed at Dr. Babasaheb Ambedkar Marathwada University, Aurangabad. We are thankful to the authorities of the libraries of Dr. Babasaheb Ambedkar Marathwada University, Aurangabad, Gokhale Institute of Politics and Economics, Pune and Marathwada Agricultural University, Parbhani for extending their kind cooperation and ready help in carrying out the present study. We have collected statistical data from the offices of the fourteen agricultural produce market committees in Marathwada. We thank all of them for providing the required data.

We would like to thank Mr. Muraleedharan for neat and prompt typing. I also thank Serials Publication, New Delhi for their keen interest in undertaking the publication of this manuscript.

S.R. Takle
V.B. Bhise

1

INTRODUCTION

IMPORTANCE OF PRICES

National economy of India is still dominated by agriculture, because agriculture commands a larger part of the total national income. There is a close relation between the agricultural sector and the whole national economy. Prices of agricultural products and particularly the foodgrain prices play an important role, not only in the agricultural sector but in the whole national economy of India. Thomes and Foote[1] say that "Economics has been defined as the price aspects of human activity". Every phase of economic life is directly related to prices.

Besides agricultural development, agricultural prices have also a vital role to play[2]. The anxiety against any abnormal rise in foodgrain prices in such an economy is clearly understandable for in such economies a majority of the population devotes a substantial part of its energies to earn income, sufficient to purchase foodgrains. Expenditure on foodgrains form the major part of total consumer expenditure, culminating in many cases to eighty percent.

Parthsarthy and Mudahar[3] have pointed out that foodgrain prices is a crucial factor shaping the growth of foodgrain production, inter-sectoral distribution of income, inter-sectoral resources transfer and overall growth of economy.

According to Acharya and Agarwal[4], agricultural prices are important for governments on several counts some of which are : (1) Overall economic control, (2) component of agricultural development strategy, (3) plan and project formulation, (4) monetary and fiscal policies, and 5) market information and intervention. Most governments keep in mind the objectives of assuring remunerative prices to farmers and making goods available to consumers at reasonable prices. To achieve these conflicting objectives, close monitoring of prices at the farm gate and retail levels becomes necessary.

According to Satish Chandra[5], prices play a very important role in agriculture. In many countries for many crops no two years are alike. The agricultural prices frequently fluctuate for various reasons. The prices suffer from violent fluctuations due to vagaries of nature whereby demand and supply variations become significantly reactive. Some times even an unimpeachable factor causes distress for sales and disturbs the price movement". Hence agriculture is not only a gamble on monsoons but also on prices.

Parthsarthy and Mudahar[6] have pointed out that the whole economy of the country is affected by agricultural prices through the impact on (1) change in foodgrains production and non-foodgrain production, (2) change in domestic terms of trade between agricultural products and non-agricultural products, (3) labour cost and its effects on overall cost structure, and (4) impact on change in the pattern of income distribution.

According to Mellor[7], agricultural prices may play a more important role in signalling the Government the need for greater investment in roads, power supply, extension, research and other parts of the infrastructure of technological change in agriculture.

According to Singh[8], agricultural prices play a dominant role in economy not only because it commands a larger share in national income but its contribution to change in overall prices is more than non-agricultural prices.

Acharya and Agarwal[9] say that agricultural prices are most important for farmers Individual farmers make use of prices in various ways: (1) Long-term planning of farm business: knowledge of prices is useful to farmers in making long term farm investment. Investment in buildings, farm machinery, land development and irrigation structures is long-term investment. Prediction of likely returns from this investment is done on the basis of prices, (2) information on prices is beneficial to farmers in taking decision pertaining to selection and management of different crops and livestock enterprises. The optimum combination of different enterprises is worked out under a limited resources situation by using the principle of opportunity cost, and (3) planning the time, place and method of marketing farm products and purchase of inputs. The price of input varies from time to time and brand to brand. The knowledge of prices helps farmers to take the correct marketing decisions consistent with the goal of profit maximization.

In the explanation of importance of agricultural prices Kahlon[10] has pointed out that in the developing countries like India, where 50 percent of national income is shared by agriculture, agricultural prices affect the general price level. On the other hand, a measure taken to control general price level has an effect on agricultural prices.

According to Acharya and Agarwal[11], agricultural prices are important for cooperatives and private business organisations, in various ways. They help them to plan the character, location, size and other features of the agricultural business enterprises. It also helps them to determine the time and place of purchase of raw materials or supplies and govern policies with respect to expansion or scaling down of the business. Prices help them in selecting the markets for their products, the timing of the sales and the adoption of promotional sales efforts.

Agricultural prices are important from the point of view of the standard of living for the various classes of society as consumers. Because change in agricultural prices have different types of effects on the standard of living of these various classes. An increase in agricultural prices brings down the standard of living of consumers.

Raj Kumar Singh[12] has stated that the agricultural prices play an important role in Indian economic planning. In democratic planning excessive control and regimentation are not possible since the methods and resources available with totalitarian planning are absent.

According to Indrasena Reddy[13], agricultural prices also serve as a guide to farmers in their decisions regarding what to produce, when to produce and how much to produce. Furthermore, prices of agricultural commodities also act as signals for the movement and allocation of resources and in turn affect agricultural production. In view of the importance of agricultural prices, there has been an increasing recognition of the need to keep a watch on their movements, which is necessary to formulate appropriate farm price policies.

Kaunal has stated that prices play an important role at every stage in production, distribution, exchange and all other activities till the agricultural produce reaches the ultimate consumers.

According to Chalam and Sanjeeva Rao[14], "in the economy, where the agricultural sector accounts for almost half of the national income, both the rise and fall of agricultural prices would have far reaching consequences on the general level of employment and income in the economy".

Agricultural prices have an important effect on government revenue, funds, taxes, etc. They also affect the standard of living of agriculturists; marketable surplus as well as earning of foreign exchange also depend on agricultural prices. If agricultural prices have increased, farmers will sell more quantity in market. It is also possible to export more quantity due to larger marketed surplus, in the international market and hence earn more foreign exchange. Therefore, it is most important to have such level and structure of agricultural prices that satisfy all these various aspects of the economy.

PRICE PROBLEM OF AGRICULTURAL PRODUCTS

Agricultural prices is a big problem in Indian agricultural sector, because agricultural production depends upon monsoons, therefore the agricultural prices are subject to fluctuations any time. There are two aspects. of agricultural price problem. The first is: Are prices of agricultural products rational and incentives? The second aspect of agricultural price problem is arbitrary variation in the prices of agricultural products.

Krishnamachari Committee[15] in its report has stated that in the absence of incentive and stable prices the producer will have to suffer the penalty of receiving uneconomic prices in return for his enterprises in increasing the productivity of his land.

Now a days price variations is a serious problem for farmers. The farmers have always been the losers and do not get the profits. The study of the second aspect of this problem, that is the problem of seasonal variation in the prices of agricultural products has been conducted by many researchers and have made the following points:

Rajbans Kaur[16] says that in the free market economy the prices of agricultural commodities are unexpected and fluctuate violently, which have adverse effects on the pattern and volume of agricultural production.

There are two types of price variation: (1) spatial price variation, and (2) temporal price variation. The spatial price variations depend on element of distance and temporal price variations are based on time element.

TYPES OF PRICE VARIATIONS

The well known feature of agricultural price variations is that the prices move downwards at a minimum level, in peak marketing period and increases to maximum level in lean season[17] But the seasonal variations are different from commodity to commodity and time to time.

There are six major time elements in price variation[18]: (1) Short-time price variations, (2) seasonal price variations, (3) annual price variations, (4) cyclical price variations, (5) secular trend in prices, and (6) irregular and episodic price fluctuations.

Short-time Price Variations

In any price series there are many minor variations, lasting comparatively short periods of time, which have a little or less regular movements of long-run duration. In other words, short-time price variations occur due to changes in the demand and supply. There are the hour to hour, day to day or week to week price variations. The main difference between agricultural prices and marketing prices of other products is the greater and more frequent fluctuations of agricultural prices. According to Thomes and Foote[19], short-time price variations are caused mainly by three factors viz., variations in receipts, temporary fluctuations in consumer demand and the experimentation involved in the discovery of supply and demand conditions and prices.

Weekly short-run price variations take place in case of perishable commodities, for example, fruits, vegetables, milk, etc. The prices of vegetables are not stable on a whole bazar day. The prices may be different in the evening from that noticed early in the morning on the bazar day in rural areas.

Seasonal Price Variations

The second distinct type of variations observed in the prices of most commodities is the seasonal movement. This type of price variation is more important and at the same time more simple and better understood than short-time price fluctuations. The seasonal nature of agricultural production, low capacity of the small and marginal farmers to hold stock and to obtain credit to postpone their sales, less storability of agricultural products are the factors which explain the variations in arrival and prices of agricultural commodities from one season to another within a year.

According to Bhise[20], the tendency is that the market supply of agricultural commodities is very high in peak period and very low in lean period. The excess market supply (arrival) of agricultural commodities pushes the prices down to very low levels in peak period and low market supply (arrival) pushes them to unduly high levels during lean period. Naturally, the peak and lean periods differ from commodity to commodity.

The seasonal nature of production affects prices of agricultural commodities in two ways[21]: (1) Perishable and semi-perishable commodities have to be moved to consumption points as soon as the produce is harvested. In that part of the year when supplies are large, there is bound to be a decline in prices and vice versa. This is true for vegetables, fruits, milk and other perishable products. (2) Non-perishable commodities like foodgrains, cotton and wool can be stored after the harvest. The storage of agricultural products involves cost called 'storage cost'. The prices of such commodities fall during the harvest season and rise thereafter in proportion to the storage cost. Inter-seasonal variations in prices of non-perishable commodities are less as compared to that in prices of perishable commodities.

Seasonal price variation of an agricultural commodity is important to producers and to those who process and market the commodity. Since the seasonal normal price movement depends solely on holding costs, elasticity of neither demand nor supply has a bearing on this price rise[22].

It is true that the agricultural products and foodgrains have a seasonal behaviour. Agricultural products which are perishable and semiperishable cannot be stored for a longer period. So the farmers sell their products just after the harvest. The supply of agricultural goods increases in harvest period and prices decrease while the supply declines in lean period and prices increase to a higher level[23].

Annual Price Variations

Changes in prices from one year to another are known as year to year or annual price variations. Such fluctuations are observed for agricultural commodities mainly because of the annual nature of production. The area under cultivation and yield of crops fluctuate from year to year. For some crops, however, acreage changes are more responsible for yearly fluctuations in production than the variations in yield.

Annual price variations result in fluctuations in the farmers' incomes. If the demand for farm products is inelastic, the annual variation in yields causes prices to fluctuate. It means that when demand for farm products is inelastic, the producers incomes will vary from year to year with crop yields[24].

Secular Trend in Prices

The tendency of prices to move up or down over a longer period of time is termed as a trend or secular price variations. A trend in prices is usually established on the basis of at least 10 to 15 years data. The trend is not concerned with the movement in prices from one year to another but for a large number of years say, ten or more.

Thomes and Foote[25] have stated that any slow changes in supply or demand which take place over a long period of time may cause a gradual but nevertheless distinct rise or fall in price. For example, the steady growth in population or the gradual development of new uses for a commodity may result in a very slight rise in price each year, in addition to the effects of the changes in demand and supply. These long time changes are referred to as 'secular trends'.

Waugh defines secular trend as irreversible movements which continue in general, in the same direction for a considerable period of time. It means the secular trend is the long period tendency of the data.

Cyclical Price Variations

Agricultural prices also change in cyclical manner but are different from seasonal variations. Cyclical price variations occur in long period than the period of occurrence of seasonal variations and they are unplanned, because they are the result of the factors which are beyond the control of farmers. The length of price cycles vary from commodity to commodity. As well as prices and productions do not move in similar cycle because sometimes change in prices and weather conditions are also not regular[26].

It is common knowledge that the price of any farm product decreases during the harvest season and continues to rise during the rest of the year till the next harvest season. Prices of most farm products also reveal a cyclical pattern within a year[27].

Irregular Price Variations

Irregular price variations are due to wars, droughts, floods, earthquakes, strikes, elections, early or late arrival of the monsoon, fear of tax rise especially at the time of the presentation of the budget by the Government or any incidental factors which are not concerned with the general trend or cyclical price variation or annual price variation. It is impossible to avoid these variations and also difficult to control them. Irregular price variations may be of shorter or longer duration.

CAUSES OF PRICE VARIATIONS

Agricultural price variations is a conspicuous phenomenon in India. The root of the kind of price variation is in the nature of demand and supply of agricultural products. Also there are number of other factors which result in fluctuations in the agricultural prices, viz., change in monetary policy, fiscal policy, tariff policy, future price trend, transportation facilities, storage and credit facilities, processing facilities, effective marketing system, economic and political events, lack of effective price policy, change in technology, credit policy, rate of taxes, government facilities, nature of production, change in market arrivals, etc.[28]. But wide fluctuations in agricultural prices take place mainly due to the difficulty of adjusting supply to change in demand for agricultural products[29]. There are various factors which are responsible to price variations of agricultural product. Some important factors causing price variations are discussed below :

Inelastic Nature of Demand

The production of agricultural commodities takes place in a specific period of year but the demand is spread all over the year. In nature the demand of agricultural commodities and mainly foodgrains demand is less elastic to change in its prices. Dandekar[30] has pointed out that 'generally low price elasticity of consumers demand for foodgrains, it needs a disproportionately large rise in prices in order to restrict the demand to a somewhat smaller harvest and it needs a disproportionately. Large fall in prices to expand the demand to a somewhat large harvest. The market prices of foodgrains are thus liable to large fluctuations from year to year, firstly because of the fluctuations in the size of the harvest and secondly because of the generally low price elasticities of demand for foodgrains. Raj Kumar Singh[31] has stated that agriculture produces in the main, foodstuffs which are the basic necessities of life. In result, the demand for all agricultural products, taken together is inclastic Land assumes as a capital sunk factor in agriculture than in any other industry. Because invested capital cannot be shifted from these to some other sector. The supply of agricultural labour is also inelastic, when the farmers work on their land themselves with their family, many of the farmers are self-sufficient in labour as a factor of production. So the amount of labour spent on cultivation hardly shows response to change in prices of agricultural products. This situation further results in increased inelasticity of supply of agricultural commodities[32].

Agriculture is a Biological Process

One important aspect which shows larger instability in agricultural sector than industrial sector is that the industrial sector cannot be affected by indigenous factors as agriculture. Mostly instability in industrial sector is attributed to certain exogenous factors, operating outside the immediate sphere of the industry concerned. But the fluctuations in agricultural prices and production have their own peculiarity. Supply of agricultural products is less sensitive to change in prices. Because agriculture is a biological process where human control is limited. It is very difficult for farmer to alter his programme of agricultural production in short period. Once the process of growing of crop is started it cannot be slowed or speeded up[33]. Agricultural prices are more sensitive than the prices of industrial products to change imbalance between supply and demand[34]. It indicates that agricultural production is uncertain which depends on monsoon and its demand is relatively stable. This situation further generates wide fluctuations in agricultural prices.

Domination of Nature

Agriculture mostly depends on nature which is out of control of human efforts. Schultz has stated that agriculture is the mercy gamble of nature. Droughts and dustwarms, floods and frosts are manifestations of the struggle that goes on forming between man and. nature. These different types of vagaries in agriculture faced by farmers, involved various forms of economic risks and uncertainties[35].

So it has been well known dictum in case of Indian agricultural problems that the vagaries of monsoon and other factors are cause to greater fluctuations in production and prices of agricultural products.

Fixed Nature of Cost of Production

The cost of production of agricultural production has a special strategy than the cost of production in industrial sector. The nature of production cost is almost fixed or contracted, for example, rent, revenue, payment of interest, and family labour. The variable cost is manures, fertilizers, labour, seed, equipment etc., which is a small percentage to total cost of agricultural production. But in case of industry the proportion of variable cost is larger. So it is easy to curtail industrial production. When prices fall, which is not so easy in case of agriculture[36]. The structure of production cost in agriculture is relatively rigid and more slowly than the level of prices especially in times of falling prices.

Combination of Different Products

Choudhary draws attention to a different type of aspect of agricultural sector. Agriculture is a many industries rolled into one. It included several industries such as cereal farming and vegetable growing, dairy farming, etc. Further, it is divided according to kinds of products grown. There are some jointly produced commodities where decision regarding prices becomes difficult Because separate cost of production cannot be shown. Many times one agricultural commodity is required for several purposes, and farmers find it difficult to anticipate market demand. It is due to this, agricultural sector becomes more unstable[37].

Seasonal Sale of Agricultural Products

A well known reason of price variations of agricultural products is a seasonal sale of products. Due to economic difficulties the farmers are compelled to sell their products in harvest period. Such sales of agricultural products by farmers are known as 'distress sales'. This leads to fall in prices of farm products during harvest period. The supply of products to the market varies from year to year; also depending upon the production during the year. Moreover, arrivals on market is only surplus production over and above the needs of products. Nearly one-half to three-fourth of the total marketed quantity is sold in four months of harvest period. Therefore the fluctuations in the marketed quantity are proportionately larger than the fluctuations in the production[38]. This is supported by the empirical study conducted by Kamala Devi[39]. In short, seasonality in sale of agricultural products and changes in marketed quantity are the factors which are responsible for occurrence of price variations.

Speculative Demand of Wholesale Traders

Dandekar explains that the speculative demand from wholesale traders is an effective factor for seasonal price variations of products. The wholesale traders buy the agricultural products from the farmers which arrive on the market and stock it. Further, the stocks are released on retail market to meet the demand of final consumers which is, more or less, steady. Demand for agricultural products from wholesale traders for stocking depends on price expectations. Hence it is speculative in nature. This speculative demand of wholesale traders is destabilising in kind. So when the arrivals on the market are larger, the prices are at lower level. "It is not difficult to see that this speculative demand of wholesale traders is essentially of destabilising in kind. When the arrivals on the market are larger and the prices are low, when arrivals are meagre and the prices are high, it pushes the prices further up"[40].

Thus the variations in the supplies of products coming to producers market combined with speculative demand of wholesale traders cause to wide fluctuations in prices of foodgrains. The same thing happens in retail markets also. Normally, the wholesale traders should release their supplies depending upon their stock and their expectations of the next and future harvests. They vary their supplies to retail market according to their future expectations. The final consumers demand which is price inelastic, even small variations in supplies bring about large fluctuations in final prices. Thus the operation of wholesale traders is a cause of variation in prices in producers' as well as in consumers' market. "The speculative demand for stock spreads to the producers on one hand and to the retailers and consumers on the other hand. This is equally destabilising and further accentuate the fluctuations in prices"[41]

Vicious Circle Between Prices and Sales

It. is an interesting but very serious aspect from the farmers point of view that there is like running race between price and sales of agricultural products. It means once agricultural prices begin to move downwards, it push sales up and further price falls down. Because due to falling trend of agricultural prices the farmers cannot save much quantity by reducing his production. But indeed the farmers attempt to increase their production and also attempt by selling larger quantity to maintain their money income. This operation push prices again to downwards. This is specially more true for the farmers carrying an agriculture with family labour as a way of life rather than a business. Thus the falling trend of agricultural prices is accompanied with increased supply because the farmers have a loss of their real incomes in falling prices. It results in fall in the standard of living and particularly for marginal farmers, it would be difficult to meet their necessary needs. The small farmers have to bear double loss due to fall in prices. Firstly they have small quantity of output and secondly they sell it at prevailing low price level. So over the range of falling prices, farmers are forced to produce more and sell more surplus to maintain their total cash income to meet their other financial requirements[42]. It means in agriculture there is the backward slopping supply curve, that is, when prices fall supply rises because farmer tries to get the same real income from the output. As the prices fall marginal utility of the last unit increases for him and farmers fall into a vicious circle. Therefore, once the prices start falling they cannot be checked easily[43].

EFFECTS OF PRICE VARIATIONS

Agricultural prices are more sensitive to change in imbalance between supply and demand than the prices of industrial products[44]. As a result, price variation take place by dimension of time, space and form[45]. It means prices fluctuate from time to time, from market to market and from commodity to commodity. There are various effects of the price variations on the national economy, such as effects on economic life of farmers, allocation of resources, production pattern, trade, land distribution, monetary policy, fiscal policy, marketing, standard of living, capital formation, etc.[46] In this connection Rogar Thomes points out that "next to the rain price changes have been the greatest enemy of the farmers". Some important effects of agricultural price variation are discussed below:

Change in Terms of Trade

The degree of price variations differ from commodity to commodity which results in change in terms of trade between foodgrains and cash crops, and between agricultural products and industrial products. Price variations also bringforth undesired changes in allocation of resources and economic imbalance in the sectors of economy. In many empirical studies it is found that the price variations in agricultural sector are greater than that in industrial sector. As a result, the terms of trade between agriculture and industry has changed unfavourably to agricultural sector. Tyagi and Kohlon have found that the terms of trade have turned against agriculture because the degree of increase in agricultural prices lagged behind the industrial prices. The agricultural prices could not keep parity with prices of manufacturing. So the industrialists became rich and on the contrary the agriculturists specially small farmers became poor[47]. Dandekar has said that in India the terms of trade between agriculture and rest of economy is not favorable to agriculture. Due to this, agricultural sector cannot attract more resources and this position of terms of trade has remained more or less unchanged from the last fifteen years[48]. This type of disparity in price variations generate sectoral imbalances in the economy.

Depress the Incentive of Farmers

Incentive of farmers to increase production is affected by variation in prices of agricultural products. If the prices of commodities vary irregularly the producers will have very little incentive to get in its production or to expand production where it is already growing. On the other hand, stable price level of a commodity provides strong incentive to producers for increasing production by expanding the area and by adopting improved farm practices. According to Adur, a steep decline in the prices bring in heavy loss to the growers of the crop. This will reduce their income and dampen their spirit to produce more next year[49].

Effects on farmers Income Level

A large number of agriculturists in our country falls in the lowest income group and hence they are indebted. Irregular fluctuations in agricultural prices have adverse effects on the income level of these agriculturists. In the harvest period they are forced to sell their product at lower prices. Therefore, their income is kept at a minimum level. On the other side, they cannot obtain the benefit of higher prices which prevail in lean period, because firstly they have very limited quantity of marketable surplus; secondly, increases in agricultural prices, mainly cereal prices raise the prices of other goods which are required to farmers for their business and to meet their needs of inputs, capital equipments, consumption articles, etc. So the terms of trade may not favour to farmers[50]. The real income of farmers is at a relatively low level and they would gain only when prices of agricultural products are kept at higher level as compared to that of other manufactured articles and agricultural inputs which they require.

Effects on Farmers' Economic Prosperity

The farmers' economic prosperity depends upon price variations of agricultural commodities produced by them. Natarajan has found that within agricultural sector there is a disparity in prosperity between growers of foodgrains and growers of commercial crops. He found the substantial increase in the prosperity of producers of commercial crops as compared to foodgrain producers[51]. It reveals that there is no price parity between the various agricultural commodities. It results into intra-sectoral economic imbalances. Further, Natarajan has pointed out that the disparities which have noticed in the prosperity of the growers of foodgrains and commercial crops is a serious problem and if allowed to continue unchecked may have serious repercussions on the progress of grow more food efforts[52].

Effects on Small Farmers

Generally the agricultural classes are benefited from the agricultural production. But in case of India where nearly 70 percent of farmers having small and marginal land are not able to consume the benefits of increased agricultural prices because they are compelled to sell their products in harvest season at lower prices. The benefits of higher prices of agricultural products are appropriated by big landlords and middlemen only. Farmers have greater loss in the falling prices sometime in the period of bumper production in agricultural sector, prices fall down suddenly below the level of cost of production also. Therefore, small farmers face the economic loss[53].

Black Marketing and Speculation

Black-marketing and speculation in various sectors of the economy is also an effect of fluctuations in agricultural prices. When the inflationary situation prevails in the market, the speculation and malpractices take place by traders and businessmen due to scarcity of supply and defective marketing control. These factors push up agricultural prices. These malpractices mostly begin in the godowns of both traders and manufacturers and not in fields of agriculturists. It indicates that black marketing and speculation are done by traders and industrialists; and benefits of these malpractices are absorbed only by them.

Lack of Regional Integration of Agricultural Markets

Due to irregular fluctuations in prices of agricultural products their price level differs from region to region in the country. It is known as regional price disparity. Sometimes the disparity in prices is greater than the cost of transportation and other handing costs. It reveals the lack of regional integration among agricultural markets. Due to this the degree of marketed surplus is irregular and hence it changes from one region to another. The benefits of regional price disparities are appropriated by big traders and businessmen. They purchase larger quantities of agricultural products in the region where the prices are lower and sell in the region where prices are higher. The export of agricultural products and ultimately earning of foreign exchange are also affected by regional disparities in agricultural prices. According to Gadgil[54], uniformity in agricultural prices is essential factor to attain an integrated development of national economy He said that the price differentiations from one region to other region should not be higher, than reasonable transportation and other costs.

Effect on Price and Production

Rise and fall in agricultural production and that in agricultural income (prices) do not synchronise. Agricultural production is a prolonged process. There is no coincidence between change in prices and change in production. This situation generates cyclical fluctuations in prices and production and is difficult for the agriculturists to break this cycle. Cohen has pointed out that the initial favourable prices stimulate an increase in output and when the resulting larger supplies come to market, prices fall The farmers decide to contract output and again after a time lag prices rise. The continued existence of this cycle besides necessitating certain relationship between the elasticities of demand and supply, assumes that farmers do not learn from his experience that there will be a cycle[55].

Change the Producer's Share in Consumer's Rupee

The share of producer in consumer's rupee is different in peak and lean seasons, due to the fluctuations in agricultural prices. It is the established fact that producer's share in consumer's rupee is low in peak marketing season and high in lean marketing season[56]. A large number of farmers sell maximum quantity in peak season when producer's share is the minimum and a smaller quantity is sold in lean season when producer's share is the maximum. It indicates that the farmers particularly small and marginal farmers have to bear financial loss in both peak and lean seasons.

Effect on Government Budget

Agricultural price variations and mainly foodgrains price variations have effects on government budget. In developing countries a significant part of budget is devoted to salaries of employees. When foodgrain prices rise, the government is forced to pay compensatory dearness allowances to its employees. So by reducing investment from productive agricultural sector, government has to divert the amount for payment of higher salaries to employees to maintain their standard of living. Further it induces to increase government expenditure and also money supply which results in inflationary trend to the economy[57].

An attempt has been made to study the seasonal, annual and long- term behaviour of prices and arrivals of foodgrains and oilseeds in Marathwada region of Maharashtra state.

OBJECTIVES

This study concentrates mainly on the following objectives:

1. To study the seasonal variation in prices and market arrivals of foodgrains and oilseeds at regulated markets.
2. To study the seasonal pattern of market arrivals and prices of foodgrains and oilseeds at regulated markets.
3. To study the annual variation in prices and market arrivals of foodgrains and oilseeds at regulated markets.
4. To study the relationship between market arrivals and prices of foodgrains and oilseeds at regulated markets.
5. To study the trends in prices and arrivals of foodgrains and oilseeds at regulated markets.

RESEARCH METHODOLOGY

(i) Study Area

Marathwada is a backward region of the Maharashtra state. The main occupation of people in this region is agriculture. 81 per cent of the total population depends on agriculture in this region. If agricultural sector is developed it would be possible to achieve the development of industrial sector as well as the whole economy of the Marathwada region.

(ii) Selection of Markets

At present there are seven districts in this region. Two marketing centres from each district of the Marathwada region are chosen for the present study. In all 14 regulated markets are selected for the purpose of the present study. While selecting the regulated marketing centres the factors like the size of markets, their integration with superior markets/other markets, size of the trade area, size of market arrivals, annual income of the markets, infrastructure available at the market places, number of registered traders/commission agents, etc. can be considered However, the income of the market committee could be taken as the best criterion for choosing the regulated markets for further studies. Government of Maharashtra has classified all the regulated markets in the state into seven groups. The grading system of the State Government for regulated markets is shown in the following table:

Table 1
Grading of Regulated Markets in Maharashtra

Sr. No.	*Grade of market*	*Annual income range*
1.	Super grade	Above 10 lakh
2.	'A' grade	3 to 10 lakh
3.	'B' grade	1.75 to 3 lakh
4.	'C' grade	1.00 to 1.75 lakh
5.	'D' grade	50,000 to 1,00,000
6.	'E' grade	40,000 to 50,000
7.	'F' grade	25,000 to 40,000

Source: Maharashtra State Agricultural Marketing Board, Annual Report of Agricultural produce market committees in Maharashtra, 1994-95.

At present there are 63 regulated marketing centres in the Marathwada region. Most of them belong to the first three upper groups. In Marathwada, the number of regulated markets belonging to super grade group is 4. The number of 'A' grade markets is 23 and there are twelve 'B' grade markets in Marathwada. The number of 'C' and 'D' grade markets is 15 and 9 respectively. Districtwise and gradewise distribution of regulated markets in Marathwada is shown in table 2 below:

Table 2
Gradewise Distribution of Regulated Markets in Marathwada

	Districts							
Grades	*Aurangabad*	*Jalna*	*Parbhani*	*Nanded*	*Osmanabad*	*Beed*	*Latur*	*Marathwada*
Super	1	1	—	—	—	1	1	4
A	3	2	7	3	2	5	1	23
B	1	1	3	4	1	—	2	12
C	2	1	1	6	2	1	2	15
D	1	—	2	2	2	1	1	9
Total	8	5	13	15	7	8	7	63

Source: Agricultural Marketing Board, Divisional Office, Aurangabad, 1996-97.

Most of the markets belonging to 'B', 'C' and 'D' grade groups have been established very recently. So the time series data on monthly arrivals and prices in case of these markets are not available for the period chosen for the present study. So the choice of the regulated markets for the present study was restricted mainly to super grade and 'A' grade markets in Marathwada For the purpose of the present study the chosen 14 regulated markets include four super grade markets viz., Aurangabad, Jalna, Latur and Manjlegaon, nine 'A' grade markets viz., Parbhani, Beed, Osmanabad, Nanded, Partur, Manwat, Kallam, Udgir, Lasur and one 'B' grade market viz., Naigaon. The districtwise list of the chosen regulated marketing centres is shown in the following table:

Table 3
Regulated Marketing Centres chosen for the Present Study

District	*Name of the market*	*Grade*
Aurangabad	Aurangabad	Super grade
	Lasur	'A' grade
Jalna	Jalna	Super grade
	Partur	'A' grade
Latur	Latur	Super grade.
	Udgir	'A' grade
Parbhani	Parbhani	'A' grade
	Manwat	'A' grade
Osmanabad	Osmanabad	'A' grade
	Kallam	'A' grade
Beed	Manjlegaon	Super grade
	Beed	'A' grade
Nanded	Nanded	'A' grade
	Naigaon	'B' grade

(iii) Selection of Commodities

The main aim of the present study is to study the behaviour of arrivals and prices of main foodgrains and oilseeds in Marathwada. This study covers nine main agricultural commodities produced in this region. The selected foodgrain commodities are wheat, Kharif jowar, rabi jowar, tur, bajra and greengram. These six foodgrains cover 50.76 percent of the gross cropped area as against 63.06 percent under all foodgrains (in 1995-96) in Marathwada. The selected oilseeds include groundnut, sunflower and safflower. These three oilseeds cover 10.81 percent of the gross cropped area as against 15.14 percent under all oilseeds (in 1995-96) in Marathwada. These foodgrains and oilseeds together account for 61.57 percent of the total gross cropped area in Marathwada during 1995-96.

After having collected the data on monthly arrivals and prices of selected agricultural commodities from the office records of the chosen regulated markets, we came to know that in case of some commodities the arrivals are very low in few markets. Hence we decided to drop such markets where arrivals are very low. All the 14 markets have been considered in case of wheat and tur. Naigaon regulated market is dropped from the analysis of rabi jowar, Manwat, Manjlegaon and Nanded markets have been dropped in case of kharif jowar while Udgir, Naigaon and Nanded markets have been dropped in case of safflower. Four markets (viz., Parbhani, Osmanabad,. Nanded and Naigaon) have been dropped in case of bajra and another four markets (viz., Aurangabad, Partur, Lasur and Naigaon) in case of groundnut. Beed, Osmanabad, Manjlegaon, Lasur and Naigaon markets have been dropped in case of greengram. Only six markets (viz., Parbhani, Beed, Manjlegaon, Kallam, Osmanabad and Latur) have been considered for studying the behaviour of arrivals and prices of sunflower.

(iv) The Data

This study covers the period of 15 years, from 1981 to 1995, and is based completely on secondary data. The time series data regarding monthly arrivals and average prices of the selected foodgrains and oilseeds are collected from the records of the selected agricultural produce market committees in Marathwada.

(v) Tools of Analysis

In the study some statistical tools have been used for analysing the data on arrivals and prices of selected agricultural commodities. These tools are as below:

Seasonal variation

The variations in the monthly market arrivals and monthly average prices of selected foodgrains and oilseeds are measured by using the method of simple averages.

Annual variation

In order to measure the year-to-year variation in annual market arrivals and annual average prices of selected foodgrains and oilseeds the method of simple averages is employed.

Range

In order to measure the seasonal and annual variations in arrivals and prices, the range between the maximum and minimum values of arrivals and prices of selected agricultural commodities is taken.

$$R_0 = R_b - R_a$$

Where R_0 = Range.

R_b = Maximum level

R_a = Minimum level

Standard Deviation (S.D.)

The technique of standard deviation is employed to measure the seasonal and annual variations in arrivals and prices of selected agricultural commodities. The standard deviation (S.D.) is calculated as below.

$$\sigma = \sqrt{\frac{\Sigma\, dx^2}{N}}$$

Where, σ = Standard deviation

dx = deviation from actual mean i.e. $(\bar{X} - X)$

$\bar{X}$ = arithmetic mean

n = Total number of observations.

Coefficient of Variation (C.V.)

Coefficient of variation is a relative measure of dispersion or variation of market arrivals and average prices of agricultural commodities It is calculated as below :

$$C.V. = \frac{\sigma}{\overline{X}} \times 100$$

Where, $C.V.$ = Coefficient of variation,

σ = standard deviation, and

$\overline{X}$ = arithmetic mean.

Coefficient of Correlation

The technique of correlation coefficient r is used to measure the degree of relationship between market arrivals and average prices of selected agricultural commodities. It is calculated as below :

$$\sigma = \frac{\Sigma\, dx\, dy}{\sqrt{\Sigma\, dx^2 \times \Sigma\, dy^2}}$$

Where, r = Correlation coefficient,

dX = deviation from actual mean of X values i.e. $(\overline{X} - X)$

dY = deviation from actual mean of Y values i.e. $(\overline{Y} - Y)$

Seasonal Index

The seasonal indices of market arrivals and average prices of selected agricultural commodities are calculated by using the method of simple averages.

$$\text{Seasonal index} = \frac{\text{Monthly Average}}{\text{Total average}} \times 100$$

Trend

The trend of arrivals of each selected agricultural commodity is studied by fitting the linear regression equation of arrivals on time with the help of least-squares method. Similarly, the trend of prices of each selected agricultural commodity is studied by fitting the linear regression equation of prices on time with the help of least-squares method. The simple linear regression equation used to estimate the annual trend values of arrivals and prices of each selected agricultural commodity for each selected regulated market is as follows :

$$Y = a + bX$$

Where, Y = Arrivals/prices,

X = Time (in years),

a = Intercept, and

b = Slope.

This book is divided into seven chapters. Chapter first deals with the importance and role of agricultural prices. It also discusses the types, causes and effects of variation in agricultural prices. The second chapter reviews the relevants literature on agricultural price variation. The third chapter presents the analysis of seasonal variation in arrivals and prices of agricultural commodities. The fourth chapter deals with the discussion on annual variation of arrivals and prices of agricultural commodities. The analysis of price trend is presented in chapter five. The sixth chapter presents the disscussion on seasonal pattern of market arrivals and prices of agricultural commodities. The seventh chapter gives the findings of the study and policy measures.

REFERENCES

1. Thomes and Foote, *Agricultural Prices,* McGraw Hill Book Company Inc., New York,. 1952, p. 1.
2. Raj Kumar Singh, *Agricultural Price Policy in India,* Printwell Publishers, Jaipur, 1990, p. 45.
3. Parthsarthy, H. and Mudahar, M.S., Foodgrain Prices and Economic Growth,. *Indian Journal of. Agricultural Economics,* Vol. XXXI, No. II, 1976, p. 16.
4. Acharya, S.S. and Agarwal, N.L.,. *Agricultural Prices,.* Oxford & IBH Publishing Co. Pvt. Ltd., New Delhi, 1994, pp. 11-12.
5. Satish Chandra,. *Agricultural Price Policy in India,.* Chug Publication, Allahabad, 1985.
6. Parthsarthy, G. and Mudahar, M.S *op. cit.*, p. 17.

7. John W. Mellor,. Functions of Agricultural Prices in Economy Development,. *Indian Journal of Agricultural Economics,* Vol. XXIII,. No. I, 1968, p. 24.
8. Singh, L.S., *Agricultural Price Policy and Stabilization Measures in India,* Capital Publishing House, Delhi, 1983.
9. Acharya S.S. and Agarwal, N.L *op. cit.*, pp. 12-13.
10. Kahlon, A.S., Stabilization of Agricultural Prices in India in the Context of Rapid Economic Development. *Indian Journal of Agricultural Economics,* Vol. XII, No. I, January-March, 1952, pp. 41-42.
11. Acharya, S.S. and. Agarwal, N.L., *op. cit.*, p. 13.
12. Raj Kumar Singh, *op. cit.*, p. 46.
13. Indrasena Reddy, Price Behaviour of Agricultural Commodities, *Indian Journal of Marketing,* Vol. 23, No. 6-7, June-July, 1995, p. 25.
14. Chalam, K.S. and Sanjeeva Rao, Agricultural Price Policy in India, in *"Agricultural Marketing : Perspective and Issues",* Arihant Publishers, Jaipur,. 1992, p. 115.
15. Kulkarni, K.R., *Agricultural Marketing in India,* The Cooperative Book Depot, Ahmed Sailar Building, Dadar, Bombay, Vol. I, 1964, pp. 519-520.
16. Rajbans Kaur, *Agricultural Price Policy in Economic Development,* Kalyani Publishers, Ludhiana, 1975, p. 1.
17. Mujumdar, N.M., Some notes on the Price Policy Implications of State Trading in Foodgrains, *Indian Journal of Agricultural Economics,* Vol. XX, No. 1,. January-March, 1965.
18. Thomes and Foote, *op. cit.*, p. 108.
19. Ibid, p. 109.
20. Bhise, V.B., Behaviour of Market Arrival and Prices, in *"Agricultural Marketing: Perspective and Issues",* Arihant Publishers, Jaipur, 1992, p. 67.
21. Acharya, S.S. and Agarwal, N.L., *op. cit.*, p. 85.
22. Warren and Trelogan, *Agricultural Market Prices,* John Wiley and East, New York, 1951.
23. Sinha, S.P., *Indian Agriculture: Its Fluctuating Fortunes,* Kitab Mahal. Pvt. Ltd., Allahabad, 1965, p. 41.
24. Bhise, V.B., *op. cit.*, p. 63.
25. Thomes and Foote, *op. cit.*, p. 117.
26. Ibid., p. 115.

27. Acharya, S.S. and Agarwal, N.L., *op. cit.*, p. 78.
28. Ibid, p. 156-157.
29. Batra, M.M., *Agricultural production,. Prices and Technology,* Allied Publishers Pvt. Ltd., New Delhi, 1978, p. 4.
30. Dandekar, V.M., Minimum Support Price for Foodgrain: Guidelines for a Policy and a Programme, *Artha-Vijanan,* Vol. II, No. 1, March, 1965, p. 274.
31. Raj Kumar Singh, *op. cit.*, p. 59.
32. Kulkarni, K.R., *op. cit.*, pp. 511-512.
33. Sinha, S.P., *op. cit.*, p. 42.
34. Mynard Geofry, *Economic Development and the Price Level,* Macmillan and Co. Ltd., London, 1962, p. 277.
35. Schultz, T.W., Economic Effects of Agricultural Programme,. *The American Economic Review,* Vol. XXX, No. V, February, 1941, p. 135.
36. Pathak, H.N., Stabilization of Agricultural Prices, *Indian Journal of Agricultural Economics,* Vol. VI, No. I,. March, 1951, p. 155.
37. Choudhary, S., Stabilization of Agricultural Prices in India in Context of Rapid Economic Development, *Indian Journal of Agricultural Economics,* Vol. XII, No. I, 1957, p. 50.
38. Dandekar, V.M., *Food and Freedom,* Dharwad University, Dharwad, 1967, p. 33-34 .
39. Kamala Devi, P., Price of Jowar in Southern zone between 1950 and 1960, *Agricultural Situation in India,* Vol. XVI, No. II, February, 1962.
40. Dandekar, V.M., *op. cit.*, p. 33-34.
41. Ibid., pp. 34-35.
42. Kulkarni, K.R., *op. cit.*, p. 512-513.
43. Choudhary, S., *op. cit.*, p. 50.
44. Maynard, Geofrey, *op. cit.*, p. 277.
45. Mcneely, John G., Some Problems of Price Analysis for Rice, *Artha-Vijanan,* Vol. I, No. 2, June, 1959, p. 128.
46. Acharya, S.S. and Agarwal, N.L., *op. cit.*, p. 158-159.
47. Adur, K.G., Price Behaviour of Agricultural Commodities, *Southern Economist,* Vol. XXIV, No. 2, 15 May, 1985, p. 12.
48. Dandekar, V.M., Agricultural Price Policy-A Critique of Dantwala, *Economic and Political Weekly,* Vol. III,. March, 1968, p. 456.
49. Adur, K.G., *op. cit.*, p. 10.

50. Sinha, Surendra Prasad and Verma, B.N., A study of Trend in Variations in Prices of Foodgrains in Bihar with Special Reference to Cereal. Prices, *Indian Journal of Agricultural Economics,* Vol. XXVI, No. IV, October-December, 1971,
51. Natarajan, B., Agricultural Price Fixation in Madras State, *Indian Journal of Agricultural Economics,* Vol. VI, No. I,. March 1951, p. 172.
52. Natarajan, B., *op. cit.*, p. 172.
53. Rajbans Kaur, *op. cit.*, p.1.
54. Gadgil, D.R., Agricultural Price Policy-A Stable for Achieving Targets. *Yojana,* Republic Number, 1961.
55. Cohen, R.L., *The Economics of Agriculture,* James Nisbet and Co. Ltd., Cambridge at the University Press, 1955,. p. 159.
56. Ministry of Food and Agriculture, Government of India, *Price Policy in India,* February, 1963, p. 114.
57. Parthsarthy, G. and Mudahar, M.S *op. cit.*, p. 18.

2

REVIEW OF LITERATURE

In this chapter an attempt is made to review the studies relating to the behaviour of arrivals and prices of agricultural commodities conducted in India. The arrivals and prices of agricultural commodities are subject to different types of variations, viz., seasonal variations, annual variations and long-term variations. A plenty of literature on these issues has appeared. This chapter deals with the important literature relating to these issues. This chapter is divided into four sections. The first section deals with the review of the studies relating to seasonal variations in arrivals and prices of agricultural commodities. Second section covers the review of the studies relating to annual variation in prices and arrivals of agricultural commodities. The studies relating to trend of prices and arrivals of farm products are reviewed in the third section, and the fourth section deals with the review of studies on the relationship between arrivals and prices of agricultural commodities.

SECTION A : REVIEW OF THE STUDIES ON SEASONAL VARIATION IN ARRIVALS AND PRICES

Kamala Devi[1] has studied the seasonality in prices of jowar for certain selected markets in the Southern zone (Madras, Mysore and Andhra Pradesh) for the period from 1950 to 1960. In this study it is found that the prices of jowar fluctuated in different extents in different markets. The prices and arrival variations between different markets centres were highest in 1956 and lowest in 1959 which indicates the different integrity of markets in different periods.

Another important finding of this study is that seasonal fluctuations in prices attained a maximum level in 1956 and prices were more stable in 1957. The price of jowar declined from November with the harvest arrivals of kharif jowar. The study also found that the degree of seasonal variations in prices is different in different selected marketing centres.

Kahlon and Dwivedi[2] have studied seasonal price fluctuations in case of wheat, paddy and cotton during the period 1951-1962 for selected markets in Punjab state. The index of seasonal price variation of wheat experienced a fall in the month of May and June with an index of 94.89 and 93.85 respectively. After June the index started rising month by month and rose to the highest level of 106.26 in February. The price movement of wheat remained less than average from April to September, when the index of price variation was less than 100. The most profitable month for marketing of wheat in the state was February.

George and Ranjit Singh[3] have studied the structure of paddy prices in Amritsar and Karnal markets in Punjab, during the period from 1957-58 to 1966-67. They have also studied seasonal pattern of arrivals and prices of paddy in Punjab. The average indices of seasonal variation for paddy prices and arrivals for the two major markets selected for this study, indicates that seasonal index of paddy price was lowest in month of November in both the markets. Thereafter it began to rise till it reached its maximum during July. Generally the price movement was characterised as less than average from September to January with an index of less than 100. For the rest of the year i.e. February to August, the index was higher than 100.

National Council of Applied Economic Research[4] studied the structure and behaviour of agricultural prices in India. The study was related to a few crops, namely, wheat, gram, jowar and paddy, which are main foodgrains in India. The indices of four selected foodgrains have been worked out for the period from 1950 to 1965. In this study seasonal price variations are explained by graphical method also. The price trends of these commodities were downward during 1950-56 and during the period from 1957-62 prices were increasing at a low rate but after 1962 prices were rising at a relatively high rate. The seasonal price variations is a common phenomenon in Indian agriculture. The range of coefficients of variation during the study period is for rice 3.1 to 10.4, wheat 2.8 to 12.1, jowar 2.8 to 20.1, and gram 4.3 to 23.8. It implies that the variation in the prices of different commodities have different degrees. The lowest price reached in the month of January when there is peak marketing season and price attained a maximum level in the month of August when market arrivals are minimum. In case of wheat prices begin to decline in or near about the month of March because of its harvest period and tends to rise during June to August. The range in seasonal index of wholesale prices for selected crops was for rice 11.6, wheat 11.1, jowar 7.8, gram 13.2. It shows that the seasonal variations are higher in case of gram and lower in case of jowar.

Seasonal pattern of market arrivals of rice showed that 50 percent to 70 percent of the total market arrivals take place between October to March. That is within six months from the beginning of harvest season, larger share of surplus grain sold in the market. During the lean season when generally prices are high, the quantity of rice marketed is a small portion of the total marketed quantity. The elasticity of market arrivals turned out to be negative. It should be noted that estimated value of coefficient of variation was very small. This means that the variations in the prices explain very little of the variations in the market arrivals during the study period.

In case of wheat 50 percent to 70 percent of total market arrivals take place during only three months. The value of correlation coefficient for the relationship between price and market arrivals of wheat turned out to be very low. It means the variation in the prices of wheat explain very little of the variations in the market arrivals. The price elasticity of market arrivals has had a negative value.

Jowar is a kharif as well as a rabi crop. The study by N.C.A.E.R. found that when coefficient of variation in prices of jowar increases, there is a general tendency of increase in coefficient of variation of market arrivals also.

In case of gram it is found that since major portion of total marketed quantity is sold during the months from April to June immediately after the harvest the value of coefficient of variation is low which indicates that the variations in prices of gram explain very little of the variations in market arrivals. The statistical relationship was found to be very weak.

Chauhan and Verma[5] have studied the price structure of bajra in Jaipur market during the period from 1959-60 to 1967-68. The seasonal variations in arrivals and prices were worked out to indicate the up and down swings that occur with some regularity during different parts of the year. This study indicated that in case of bajra arrivals rise and reach the maximum in December-January. After January, arrivals decline month by month till they almost dry up in September. Against this pattern of arrivals, the examination of the behaviour of prices shows that prices start falling as soon as the new crop reaches the market in the month of December. This pattern continues till the month of March during which the prices reach the lowest. From the month of April prices start rising again till September. Prices are lower during the summer as compared to December-January; in spite of the comparatively low arrivals during these months. This is because bajra is not usually used for human consumption during the summer. Prices start rising again from April till September because of the low arrivals and demand for seed.

To examine the variations in the price of gram, Gill and Johl[6] undertook a study for selected markets Sirsa as a big market and 7 other markets from the districts Hissar and Bhatinda in Punjab State. The period of study was from 1950-51 to 1965-66. They studied the variation in prices and arrivals within a year. In this study it was found that the price index was minimum in the month of June (92.59) and maximum in the month of January (107.89). The study also concluded that there was price differentiation in different markets due to degree of market competition. In the Sirsa market price of gram was higher because the degree of competition was high.

Bhatia and Sukheja[7] have studied the behaviour of paddy and rice prices in Rudhrapur Mandi during the period 1964 to 1969. In this study it was found that in the months of November to February, the arrivals were large as compared to other months, whereas the prices during these months were lower as compared to other months. The increase in supply (arrivals) results in decrease in price and vice-versa. For example, the prices and arrivals of grade second in the month of November were Rs. 54.88 per quintal and 21330 quintal respectively against Rs. 74.47 per quintal and 1411 quintal in August. The correlation coefficient between monthly arrivals and prices for grade third paddy was found to be $r = -0.83$. The range of prices for grade second paddy was minimum at Rs. 54.88 in the month of November to a maximum of Rs. 77.37 in the month of July.

Balwinder Singh and Sidhu[8] have studied the seasonal pattern of market arrivals and prices of foodgrains in selected markets in Punjab for the period from 1964 to 1969. They found that the price in the peak season ruled low in the small, medium and large markets alike due to heavy arrivals. The prices of these foodgrains (except maize) were generally higher in the large markets as compared to the small markets but in the medium markets the prices were almost equal to those in the small markets. The average prices during the peak marketing season for wheat, gram, paddy and maize were Rs. 68.02, Rs. 70.37, Rs. 43.12 and Rs. 55.62 per quintal respectively. In the mid season, their prices increased to Rs. 70.69, Rs. 76.57, Rs. 45.17 and Rs. 60.07 per quintal respectively. Similarly, in the lean period, the prices of these foodgrains jumped to Rs. 78.68, Rs. 77.19, Rs. 48.68 and Rs. 64.34 per quintal respectively, with a decrease in the arrival of these commodities. However, the seasonal rise and fall in the prices during the recent years in the case of wheat was not as high as it used to be before the state trading of wheat. The range of desi wheat prices, on an average, was of the order of Rs. 23.10 per quintal during the peak and lean seasons in large markets during 1959 to 1964 This range declined to Rs. 15.50 for desi wheat and for Mexican wheat the range stood at Rs. 13.23 per quintal for the period 1964 to 1969.

Sharma and Nandal[9] have studied the market pattern of seasonal index of gram in Hissar and Uklana markets during the period 1961-62 to 1972-73. There was a slight change in the marketing pattern of gram. The arrivals during the peak period increased by 12.7 percent in Hissar market and by 9.6 percent in Uklana market causing further gluts in the markets during the post-harvest period. The arrivals declined by 13.7 percent in Hissar and by 4.0 percent in Uklana markets during the medium marketing period. In the lean period the arrivals in Hissar market increased by 5.6 percent in recent five years as compared to initial years. Gram prices tended to be higher in the lean period but there was no persistent price differential in the post-harvest and lean periods.

This study indicated that the seasonal indices of gram prices and arrivals were the highest in May followed by June and April respectively in both the markets. The index of arrivals for these months ranged between 161.6 to 245.2 in Hissar and between 187.5 to 376.1 in Uklana market while the price indices during this period were the lowest ranging between 89.0 and 97.1 in Hissar and between 89.0 and 99.3 in Uklana market. On the other hand, the arrival indices from July to March ranged between 41.7 and 117.7 in Hissar and between 18.5 to 108.5 in Uklana market. The price indices for these months were higher and ranged between 94.9 and 108.7 in Hissar and between 91.1 and 109.9 in Uklana markets. Thus the major portion of the marketable surplus brought by the farmers from April to June was sold at lowest price.

Antani and Acharya[10] have studied the behaviour of market arrivals and prices of wheat and gram in Ganganagar and Madangani markets in Rajasthan during the year 1967-68 and 1975-76. In 1967-68, market arrivals of wheat in Ganganagar market were maximum during July and September and minimum in April, May and January. While in 1975-76, market arrivals were maximum in May, June and July and were minimum in the first fortnight of August and March the periods immediately following the peak market arrivals and preceding the harvest of the crop respectively. In Madangani market supply of wheat was maximum (during 1967-68) in May and June and there was no supply in February and early March. During 1975-76, maximum market supply was recorded during the middle of April and by the end of June. In case of gram market supply in Ganganagar was maximum from April end to early June in 1967-68, which is the post-harvest period, but in Madangani, market arrivals were maximum during mid-April to early May.

Kainth and Balbir Singh[11] have examined the price structure of wheat and rice at selected markets of India during the period 1964-65 to 1977-78. In this study it was found that the prices of whcat begin to decline as the harvest of crop approached or nearly took place. The index of seasonal prices fell immediately from 108.74 recorded in February to 103.75 in March. The lowest index of seasonal price was found from the month of June and was estimated at 95.38; when the marketing season was at its peak. The off-seasonal maximum index of price variation was recorded at 108.74 in the month of February. The price movements were characterised as less than average (100.00) for the months of April to October. The seasonal high as a percentage of seasonal low in respect of wheat was worked out at 14.03.

In respect of rice the study showed that the lowest seasonal price index was obtained in the month of September (89.00) while the highest index was estimated at 107.47 obtained in the month of June. The price movements were less than average for the months of October to December. The seasonal high as a percentage of seasonal low was to the tune of 20.75. Thus the study brought out that in general the off-seasonal rise in prices of wheat and rice was not very high.

Gangwar, Shrinivas and Rai[12] have examined the marketing pattern of gram in selected markets of Haryana during the period 1970-71 to 1979-80. They studied the change in the marketing pattern of gram in Dadri and Sirsa markets. They found that the sales during the lean period showed a decrease and during post-harvest period there was an increase in the sales to the tune of 5.76 and 2.24 percent and sales during mid-year period showed an increase of 3.20 and 6.74 percent in Dadri and Sirsa markets respectively. In Bhiwani market also the arrivals show an increase of 3.76 percent in peak period and a decrease of 3.11 percent in lean period. The mid-year period showed a decrease of 0.65 percent in arrivals. In the Fatehabad and Hanisi markets the peak period arrivals decreased by 3.70 and 5.67 percent respectively. The lean period sales increased by 3.42 percent in Fatehabad market and decreased by 1.00 percent in Hanisi market.

Hosamani, Hugar and Hiremath[13] have studied the seasonal pace and pattern of market arrivals and prices of cotton in two markets viz., Saundatti and Bailhongal, in Belgaum district. The necessary data were collected from the Agricultural Produce Market Committee of the two markets in regard to monthly prices and arrivals prevailing in the markets during the preceding 10 years for Jayadhar (medium staple) cotton in Bailhongal market and for 7 years for Varalaxmi (long staple) cotton in Saundatti market. They found that the coefficients of variation for both arrivals and prices in short season were lower in Saundatti market, compared to those in Bailhongal market. In the long-period, however the coefficient of variation in arrivals in Saundatti market was found to be much higher (71.48 percent) compared to that in Bailhongal market (28.73 percent). The coefficient of variation in prices was by contrast, found to be lower in Saundatti market (25.74 percent) compared to that in Bailhongal market (28.07 percent) indicating greater impact of factors other than arrivals on prices prevailing in two markets.

A study by the Reserve Bank of India[14] for the period from 1951-52 to 1964-65 observed that seasonal fluctuations in prices are wider for foodgrains than for non-foodgrains. For the entire period there emerged a tendency for seasonal fluctuations to gradually narrowing down.

Kainth[15] studied the variations in the monthly prices of wheat and paddy at Amritsar, Taran Taran and Ajanala markets and compared them between the selected foodgrains namely wheat and paddy during the period 1971-72 to 1978-79. This study indicated that the index of seasonal price variation of wheat was lowest in the month of May (90.31) followed by June (93.66). Then it began to firm up month by month and reached its peak in the month of February at 107.98. The price movement was characterised as less than average from April to October when the index of price variation was less than 100. From November to March, this index was higher than 100. The farmers can use such a seasonal index of wheat prices to market their produce on a profitable basis. The index of seasonal price variation of paddy was lowest in the month of October (89.62) followed by November (92.80). Then it began to firm up month by month and reached its peak in the month of September (108.33). The price movement was characterised as less than average from September to December, March and May. For the rest of the year the index was higher than 100.

Raj Kumar Singh[16] has studied the seasonal price indices of rice and wheat during the year 1974-85. This study indicated that the seasonal price index of rice was at its highest at 104.1 in August and at its lowest at 95.0 in March. The price of rice started falling from September on the event of approaching new crop of rice. Though it reached its bottom in March it was virtually at bottom in January-February. Thereafter it gradually rose reaching its maximum in August.

Further, during the year 1974-75, the peak price of rice in August exceeded the yearly average by 4.1 percent. And the rice price in March was 5 percent less than the corresponding averages. The price of rice was at its off-seasonal maximum in August and the marketing season lowest price of rice was seen in March, though near bottom price of rice reached in January-February.

The seasonal index of wheat at 104.6 was the highest in February and it was the lowest in May over the period 1974-85. This indicates that the price of wheat started falling from March, reached near about its bottom at 104.1 in January and at the off-seasonal trough in February coinciding. Further, during the year 1974-75 the peak price of wheat in February exceeded the yearly average by 4.6 percent. And the trough price in May was 6.3 percent less than the corresponding yearly average. The price of wheat was at its off-seasonal maximum in February and the marketing season lowest price of wheat was seen in May.

Bhise[17] has pointed out that the seasonal nature of agricultural production, low capacity of the small and marginal farmers to hold stocks and to obtain credit to postpone their sales, less storability of agricultural products are the factors which explain the variation in arrivals and prices of agricultural commodities from one season to another within a year. His study indicated that. the market supply of agricultural commodities is very high in peak season and low in lean season. The excess market supply (arrival) of agricultural commodities push the prices down to low levels in peak season and low arrival unduly push the prices to high levels during lean season. Generally the peak and lean seasons differed from commodity to commodity. This study was conducted for regulated markets in Nanded district. He found that the proportion of market arrivals of foodgrains appeared in the market was 53.86 percent in peak period and 20.54 percent in lean period. In case of pulses the proportions of arrivals appeared during these periods were 78.05 percent and 9.89 percent respectively. In case of oilseeds it was 61.41 percent and 17.89 percent respectively. It was also found that the farmers were selling larger part of their surplus output during peak period at lower prices.

Rao[18] has studied the movement in the monthly market arrivals and prices of vegetables in Pune regulated market during the period 1987 to 1992. Lady's finger experienced the highest price of Rs. 1150 per quintal in January and the lowest price in September. Similarly in the case of tomato, the higher price is noted in July, for cucumber it is January and for bottle guard it is April. Among the vegetables lady's finger noticed highest price and bottle gourd the lowest. Variations in prices are invariably less than variations in market arrivals. As of arrivals, cucumber showed a high degree of heterogeneity followed by lady's finger, tomato and bottle gourd. When prices are compared tomato has highest homogeneity and the least for bottle gourd.

Indrasena Reddy[19] has examined the seasonal index of selected agricultural commodities in Andhra Pradesh during the period 1984-85 to 1993-94. The seasonal price variation of paddy experienced a short fall in the month of January with an index of 92.61 and the low level of price continued for five months (January-May) period. However for the remaining seven-month period (June-December) seasonal indices were above average which showed that high prices were offered during this period. In case of maize, highest price index was observed in the month of August (118.44). The low price indices were noticed for the eight months period of September to March and also for the month of May. However, fluctuations in seasonal prices of maize to be more irregular than other commodities. For groundnut, low seasonal price index (93.36) was observed in the month of March and high seasonal price index (106.56) in the month of July. During the six months period from June to November, the seasonal indices are higher than average and in the rest of the months, the indices are lower than average. In case of chillies, seasonal price indices are greater than average in six months, seasonal low price (85.93) is noticed for the month of February and seasonal high price (109.99) in the month of September. Thus the analysis revealed that there are wide variations in agricultural prices in different seasons.

SECTION B: REVIEW OF THE STUDIES ON ANNUAL VARIATIONS IN ARRIVALS AND PRICES

Singh[20] has conducted a study to examine the nature of groundnut and jowar prices for the period 1953-54 to 1962-63 for selected markets in India. The author found that during the study period the range of variation between maximum and minimum yearly price index was 56 for jowar and 71 for groundnut. The author pointed out that the extent of annual price variation is greater for jowar and groundnuts.

Commission for Agricultural Costs and Prices[21] has studied the price behaviour of Kharif crop economy during the period from 1977-

78 to 1983-84. The price of commodities other than agricultural commodities as a group rose over the preceding year by 21.5, 21.8, 7.9, 1.3 and 9.2 percent respectively, whereas those of agricultural commodities rose by 9.8, 11.6, 12.4 , 5.0 and 13.7 percent respectively. Taking the entire period from 1977-78 to 1983-84, it was observed that the price of agricultural commodities had risen by 61.4 percent whereas all other commodity price had increased by 69.4 percent. In the mean time the price index of all manufactured products had increased by 62.29 percent.

Bhise[22] has found that the market arrivals and prices of agricultural commodities vary from year to year over a long period. This study clearly indicates that the variation in yearly market arrival is much higher than the variation in yearly prices. This picture is found for all agricultural commodities considered and all markets in Nanded district. In the case of kharif jowar, an average picture is that the variation in arrival is four times higher than the variation in its prices, which is the highest difference between the magnitude of variation in yearly arrivals and prices amongst the considered agricultural commodities. This fact is indicated by the coefficient of variation in arrival of kharif jowar (which is 90.92 percent) and coefficient of variation in its prices (which is 21.21 percent). On the contrary, this difference is found to be the lowest in case of gur, the coefficients being 62.5 percent and 47.14 percent respectively .

Katre, Thakare and Borude[23] have studied the variations in the annual prices and annual arrivals of paddy in Maharashtra State during the period 1974-75 to 1988-89. In this study the coefficient of variation of annual weighed prices of paddy was found to be minimum in Murbad market (17.28 percent) and maximum in Nasik market (28.66 percent). The coefficient of variation of annual arrivals of paddy was found to be minimum in Ghoti market (12.80 percent) and maximum in Gondia market (60.88 percent). It was further observed that annual arrivals of paddy were more inconsistent in markets of Vidarbha region followed by markets of Konkan region. The minimum variability in prices in different markets ranged between 2.94 percent to 5.53 percent and the maximum variability was between 9.23 percent to 17.30 percent. Regarding arrivals the range of minimum variability in different markets was observed from

28.21 percent to 98.68 percent and the range of maximum variability was from 62.42 percent to 271.79 percent.

Brahm Prakash and Sushila Srivastava[24] have conducted a study to analyse the inter and intra-year variations in market arrivals and prices of gram in all the regulated markets in Uttar Pradesh for the period of ten years. In this study the negative coefficient of correlation (-0.63) was observed between production and market arrivals of gram. The annual variability in prices was more than that of arrivals of gram. Inter-year variations in market arrivals varied between 80.1 and 130.1 percent while intra-year variations ranged from 40.2 to 193.9 percent from the average arrival during the last ten years. Above 40 percent of the total arrivals reached the market during the month of April to June when prices were low. Inter-year variations in wholesale market prices varied between 51.5 and 143.1 percent from the average annual prices that prevailed during the last ten years, while the intra-year variations ranged between 87.8 and 106.5 percent. The prices received by the farmers who sold their gram produce through regulated markets of state were more by 218 percent in 1988-89 than the procurement prices in the respective years. Moderate to high of negative coefficients of correlation were observed between market arrivals and price except for 1986-87.

Brahm Prakash and Sushila Srivastava[25] have also studied the inter and intra-year fluctuations in market arrivals and monthly wholesale prices of gram in Uttar Pradesh. The inter-year variations in market arrivals of gram varied between 86.6 and 120.0 percent from the annual average arrivals for the period 1987-88 to 1991-1992. Largest market arrivals of gram were recorded during 1988-89 while there was maximum production of gram in the state during that year. Only during 1987-88 more than 47 percent of total production of gram in the state was sold through regulated markets of the state. In the other years a low percentage of the production of gram reached the market. The wholesale prices of gram varied between 65.8 and 115.5 percent from the average price prevailed during 1987-88 to 1991-92. Except 1991-92 there was a clear upward trend. The prices of gram reached the peak level during 1990-91. The prices of gram recorded 175 percent higher during 1990-91 as compared to price of gram during 1987-88. This continuous hike in the prices during

1987-88 to 1990-91 was attributed to increase in demand for gram and general inflationary trend of the economy.

SECTION C: REVIEW OF THE STUDIES ON TRENDS IN ARRIVALS AND PRICES

A study conducted by George and Ranjit Singh[26] about the structure of paddy prices in Amritsar and Karnal market during the period 1957-58 to 1966-67 reveals that there was no visible trend, indicating stable market conditions for paddy. However since 1963-64, there has been an upward movement which is consistent with the general inflationary trend in the economy. Similarly, the pattern of market arrivals indicated a regular upward trend in both these markets. It was found that the trend in prices was much subdued as compared to the trend of arrivals.

George and Siraj Kunju[27] have examined the trend of rubber prices in Kerala during the period 1961 to 1978. The trend was not steady during the whole period. Violent fluctuations occurred from June 1965 to December 1966 as compared to the previous period when prices were unusually stable even though the maximum price fixed had been removed then. From January 1967 onwards the prices came down and this trend was visible till May 1968. Upward trend was again observed from June 1968 to March 1969. During the period April 1969 to August 1971 prices showed a gradually decreasing trend. Again from September 1971 prices were steadily increasing and this trend remained till September 1974. Since then the prices fluctuated downwards till September 1976. The situation however slightly improved from January 1977 and attained the all time high level of price during the period July-September 1978. The linear trend of prices during 1961-77 showed an average rate of increase of Rs. 24.89 per year.

A study by Adur[28] on price trend and price variations of agricultural products of cereals, pulses and oilseeds at all India level for the period from 1950-51 to 1982-1983 revealed that except for a few years since the beginning of planning in 1950-51 there has been almost a continuous upward trend in agricultural prices. The same upward trend may be seen in each sub-period and also for individual commodity group, viz, cereals, pulses, oilseeds, etc. The extent of price rise was found to be different in different sub-periods. In the first sub-period (1950-51 to 1961-62) price index increased from 110 to 123.8. In the second sub-period (1962-63 to 1970-71) it increased from 102.3 to 201.4 and in the third sub-period (1970-71 to 1982-83) from 100.4 to 248.1. Over the entire period covered under this study the price index of agricultural commodities has increased from 100 to 525.7. It means that the price has increased by five times in the period of three decades. During the first decade (1950-51 to 1959-60) price index increased by 6 percent; by 82.2 percent during second decade (1960-61 to 1969-70) and by 148 percent during the third decade (1970-71 to 1981-83).

Singh[29] has examined the relative variability in harvest prices during ten-year period (1953-54 to 1961-62) by working out the coefficient of variation. The coefficient of variation in prices was 12.5, 17.8, 17.1, 11.6 and 20.3 for rice, jowar, gur, cotton and groundnut respectively. Jowar prices did not reveal wide fluctuations within a market nor did the degree of variation differ greatly from market to market.

Antani[30] has estimated the linear trend in annual wholesale prices for cereal and pulse crops of Rajasthan during 1961 to 1975 and found that wholesale prices of all the six foodgrains increased significantly during the study period. On an average, the wholesale price index of wheat, barley, bajra, cereals, gram and pulses increased by about 17.61, 19.45, 18.74, 19.09, 33.87 and 26.26 percentage points per year, respectively. Antani's study also revealed that the rate of linear increase in prices was more than twice during 1970-75 (40.2 to 82.0 percentage points per year). Among all foodgrains, rate of increase in prices was the highest for gram and lowest for wheat in the seventies. The difference in the rate of increase between the sixties and the seventies was more pronounced for bajra and least for wheat.

Shukla and Misra[31] have concluded on the basis of the analysis of growth trend of wholesale prices of food and non-food crops during the plan periods in Uttar Pradesh that the price decreased for all crops during the first plan period but was significantly only for rice (8.6 percent), maize (4.1 percent), arhar (7 percent), castor seed (1.5 percent) and groundnut (8.6 percent). There was further decrease during the second plan period for crops like wheat, barley and jowar among the cereals and for mustard, sugarcane and potato among the non-food crops. After this period there was a rising trend and price rose by more than 12 percent per annum up to 1973-74.

Gupta[32] has studied year to year variations in harvest and wholesale prices of maize during 1952-1965 in Madhya Pradesh. The trend in harvest prices was found to be increasing while that in wholesale prices was uneven. Fluctuations tended to be widened in one market but narrowed down in the other markets.

Kainth[33] has made the analysis of trend in production, arrivals and prices of paddy and wheat in Amritsar for the period 1971-72 to 1978-79. This study indicated that the production of wheat remained fluctuated up to 1974-75. Thereafter the production of wheat showed an upward trend. The analysis brought out that the prices showed a highly significant increase (significant at 0.01 level) while the production increase was significant only at 0.10 level. The arrival and production have showed an irregular movement for the study period. However, the general trend was upward. This study revealed that the production of paddy declined upto 1973-74 and afterwards, the index of paddy production showed steep increase. Arrivals and prices of paddy have fluctuated during the study period. The analysis brought out that the production and market arrivals of paddy showed a highly significant increase. The trend in the prices of paddy was found to be non-significant. The production, arrivals and prices showed an irregular movement. However, the general trend was upward.

Raj Kumar Singh[34] has studied agricultural price behaviour during the years 1974-85. The price movements of agricultural commodities, foodgrains, cereals and pulses were examined. The magnitude of trend value for agricultural price continually increased from 136.8 in 1974-75, 194.4 in 1978-79, 238.3 in 1981-82 and 281.8 in 1984-85. Similarly, the trend value for foodgrain prices continued increasing from 151.8 in 1974-75, 197.9 in 1978-79, 232.4 in 1981-82 and 267.0 in 1984-85. Cereals also revealed the same position. Their prices showed continuously rising trend value from 150.8 in 1974-75 to 187.6 in 1978-79, 215.2 in 1981-82 and 242.8 in 1984-85. In the case of pulses also, the corresponding trend value figures were 222.9 in 1974-75, 262.1 in 1978-79, 291.5 in 1981-82 and 320.9 in 1984-85. The long-term marginal price increase during the period 1974-85 was found to be 14.5 for agricultural commodities, 11.5 for foodgrains, 9.2 for cereals and 9.8 for pulses.

Katre, Thakare and Borude[35] have studied linear trend for annual average prices and annual total arrivals of paddy in selected markets in Maharashtra state during the period from 1974-75 to 1988-89. In this study it was found that the annual average prices of paddy increased significantly at a minimum rate of Rs. 7.72 per quintal in Pen market to a maximum rate of Rs. 19.52 per quintal in Nasik market. In general, the trends in paddy prices were comparatively higher in Nasik region followed by paddy market of Vidarbha region and then Konkan region. The annual total arrivals of paddy were increasing over the time in all the selected markets except Murbad and Nasik markets. The annual rate of increase was the highest at about 75.66 quintals per year in Gondia market and lowest at about 581 quintals in Ghoti market. In general, the markets of Vidarbha region witnessed higher increase in annual arrivals as compared to the markets of Konkan region and Nasik region. Murbad market of Konkan region and Nasik market of Nasik region indicated a negative trend coefficient of annual arrivals.

Babar and Lohar[36] have examined the trends in arrivals and prices of jaggery in Sangli regulated market during the period from 1980-81 to 1991-92. The trend values of arrival and prices of jaggery showed an increasing trend over the period of years. The actual arrival of jaggery were served almost similar from the year 1986-87 to 1991-92. As regards the trend values of prices, they also showed an increasing trend as that to the arrivals.

Indrasena Reddy[37] has studied the trend in prices of paddy, maize, groundnut, chillies. This study indicated that the wholesale prices of selected commodities showed a rising tendency. The annual average price of paddy per quintal increased from Rs. 279.79 in 1984-85 to Rs..384.09 in 1993-94. The price of maize increased from Rs. 150 per quintal to Rs. 281.88, groundnut from Rs. 459.69 to Rs. 911.43 and that of chillies increased from Rs. 900.61 to Rs. 1498.63 during the same period. The price indices showed that there are fluctuations in agricultural prices from year to year and the percentage increase in the prices during the period from 1984-85 to 1993-94 is 40.80 percent for paddy, 87.92 percent for maize, 98.27 percent for groundnut and 66.40 percent for chillies. More percentage increase is observed in case of the prices of paddy. Finally, he ascertained that the increasing trend in prices of agricultural commodities indicates that the demand for these products is rising at a faster rate than the supply.

SECTION D : REVIEW OF THE STUDIES ON RELATIONSHIP BETWEEN MARKET ARRIVALS AND PRICES

George and Ranjit Singh[38] have studied the structure of paddy prices in Amritsar and Karnal markets in Punjab during the period 1957-58 to 1966-67. The relationship between arrivals and prices of paddy was worked out. It was found that the index of arrival started rising from September and reached its peak in November, when the index of prices was at its lowest. They have asserted that the relatively low variation in prices despite large variation in arrivals may be explained by the statutory fixation of prices. Since only ceiling prices were fixed, it prevented prices from shooting up during the off-season, when arrivals were low. On the other hand, when arrivals were heavy during post -harvest months, prices could not have fallen as the statutory fixed ceiling price was lower than the normal free market prices.

Chauhan and Verma[39] have studied the price structure of bajra in Jaipur market during the period 1950-60 to 1967-68. They have calculated correlation coefficients to study the relationship between arrivals and prices of bajra. In this study it was found that the prices were negatively correlated with arrival of bajra. The range of variation from average monthly arrivals was as high as between - 100 and + 181.48 percent while variation in the monthly wholesale prices from the average was 16.55 to 15.93. The relationship between yearly prices and arrivals over time was also studied. In this connection it was found that during some years the prices and arrivals were negatively correlated, while during others the correlation was positive. The simple correlation coefficient between arrivals and prices was .271 (not significant). The positive value of 'r' indicates that the increased demand of bajra did not only counter balance the effect of arrivals on prices but also induced the consumers to bid comparatively higher prices. This study also indicated that the yearly variations in the arrivals were comparatively lesser than the monthly fluctuations while the yearly variations in prices were much greater.

Verma and Falor[40] have studied the behaviour of prices and arrivals of apples in Ajmer and Jaipur markets during the year 1974-

75. In this study the relationship between arrivals and prices of apples was found to be negative. The onset of apple marketing season in the month of July, prices went down steadily with the increasing trend in arrivals. Prices were found to have reached to the minimum level in the month of September when the arrivals were maximum. From September onwards prices were found to have increased continuously with the decline in the arrivals. Prices were found to be maximum in the month of May when the arrivals were at lowest.

Jairth and Kansal[41] have studied the price behaviour of Himachal and Kashmir apples in Delhi market. They also found the negative value of regression coefficient which indicates inverse relation between prices and arrivals of apples. The value of coefficient of determination indicated that 51 percent of variations in case of Himachal Pradesh apple prices are caused by arrivals. However, in case of Jammu and Kashmir apples, the supply of apples play a very significant role. In case of Himachal Pradesh apples, prices were high when arrivals were poor and prices decline due to larger arrivals.

Chatha and Sidhu[42] have examined the relationship of arrival pattern and prices of Kharif pulses in Punjab during the period 1972-73 to 1981-82. To ascertain the relationship of arrivals and prices, correlation coefficients of monthly arrivals and prices of the last seven years (1975-76 to 1981-82) for both the mash and moong were worked out. The value of correlation coefficient worked out to be -0.7761 in case of mash which confirmed to view that more arrivals depress the market price and vice-versa. But the value of correlation coefficient in case of moong was only 0.04 as there are two crops of moong during a year. As such its arrivals were almost uniformly distributed and there were less variations over different periods and hence the low value of correlation coefficient.

Hosamani, Hugar and Hiremath[43] have studied the relationship between arrivals and prices of cotton in two important markets of Belgaum district in Karnataka state. They found statistically significant negative correlation (-0.81) between monthly indices of prices and arrivals in Bailhongal market for the years under review, by contrast the relationship between arrivals and prices in Saundatti market was found to be positive, though it was statistically non-significant (+.15).

As far as long-term relationship is concerned it was found that, in most of the months, the increased market arrivals depressed prices in both the markets, while decreased arrivals led to increased prices thereby establishing a negative association between arrivals and prices in each of the two selected markets. The coefficient of correlation for the peak period was -0.66 for Saundatti market, compared to -0.80 for Bailhongal markets indicating a greater impact of market arrivals on prices of cotton in Bailhongal market relative to that in Saundatti market for the period of peak arrivals. During the lean months, while the coefficient of correlation for Bailhongal market was estimated at -0.79, in the case of Saundatti market, it surprisingly came to 0.58. This indicates greater response of prices to arrivals in Bailhongal market compared to Saundatti market.

Bhise[44] studied the relationship between arrivals and prices of agricultural commodities in Nanded district. Correlation coefficients between monthly arrivals and monthly average prices were calculated. This study indicated that in the case of wheat, kharif jowar, paddy and gur, their monthly arrivals and prices were negatively correlated (-0.70, -0.61, -0.82, -0.91 respectively) with each other, in almost all the regulated markets. The same result is also found in the case of gram, tur, groundnut and chillies. It means that the arrivals has a negative relationship with prices in the short-run. This may be due to the fact that the maximum arrival pushes the prices down in the peak season and in the lean season minimum arrival pushes the prices to unduly high levels. A positive relationship is also found in case of some agricultural commodities at few regulated markets. This may be due to the fact that the higher prices attract the farmers and merchants to sell their stocks. The excess demand over supply may also contribute to such a positive relationship between arrival and prices in the short-run.

Rao[45] has studied the relationship between market arrivals and prices of selected vegetables during the period 1987 to 1992 at Pune market. Results showed a negative and highly significant relationship between market arrivals and prices of vegetables. An increase of one quintal of lady's finger reduced its price by about Rs. 16, it is Rs. 6 for cucumber and Rs. 9 for bottle gourd. The estimated coefficients of determination ranged between 0.23 and 0.55.

Brahm Prakash and Sushila Srivastava[46] have studied the relationship between market arrivals and prices of gram in Uttar Pradesh during the period 1987-88 to 1991-92. In this study the coefficients of correlation between monthly market arrivals and monthly average wholesale market prices of gram were found to be negative. It means that the increase in arrivals leads to reduction in prices and decrease in arrivals of gram leads to increase in its prices. Moderate degree of negative correlation was observed during all the years of this study.

The literature on seasonal variation in arrivals and prices of agricultural commodities has indicated that prices and arrivals are subject to variation from one season to another and from market to market. Further, the degree of seasonal variation also differs from commodity to commodity and from market to market. The literature revealed that the factors like the changes in demand for agricultural commodities and crop yields, explain the annual variations in both arrivals and prices of agricultural commodities. The literature pertaining to the trend of arrivals and prices has indicated that the trend in arrivals was positive in case of some commodities and negative in case of other commodities but the prices of all agricultural commodities have shown a rising trend. The literature pertaining to the relationship between prices and arrivals of agricultural commodities has indicated that generally arrivals and prices vary in opposite directions. Farmers usually sell out a major part of their marketed surplus during the peak season when prices are at their lowest level.

REFERENCES

1. Kamala Devi, P.,. Price of Jowar in Southern Zone Between 1950 and 1960, *Agricultural Situation in India,* Vol. XVI, No. 2, February, 1962.
2. Kahlon, A.S. and Dwivedi, H.N., Seasonal Price Fluctuations in Agricultural Produce,. *Agricultural Marketing,* Vol. VIII, No. 1, April, 1965.
3. George, M.V. and Ranjit Singh, Structure of Paddy Prices in Punjab, *Agricultural Marketing,* Vol. Nil, October, 1968, p . 3.
4. National Council of Applied Economic Research, *Structure and Behaviour of Prices of Foodgrains,* New Delhi, 1969.

5. Chauhan, K.K.S. and Verma, R.C., Price Structure of Bajra in Jaipur, *Agricultural Marketing,* Vol. XII, No.3, October, 1969, p. 18.
6. Gill, K.S. and Johl, S.S., Gram Price Structure in Punjab, *Agricultural Situation in India,* Vol. XXV, No. 7, October, 1970.
7. Bhatia, M.S. and Sukheja, A.L., Behaviour of Paddy and Rice Prices - A Case Study, *Agricultural Marketing,* Vol. XIV, No. 1, April,. 1971, p. 12.
8. Balwinder Singh and Sidhu, D.S., Pattern of Markets Arrivals and Prices of Foodgrains in Punjab,. *Agricultural Marketing,* Vol. XIV, No. 4, January, 1972.
9. Sharma, R.K. and Nandal, D.S., Structure of Arrivals and Prices of Gram in Hissar District of Haryana, *Agricultural Marketing,* Vol. XVII, No. 2, July, 1974, p.10.
10. Antani, K.L. and Acharya, S.S., The Change in Behaviour of Market Arrivals and Their Effect on Prices of Wheat and Gram in Two Markets of Rajasthan, *Agricultural Marketing,* Vol. XXI, No. 4, January, 1979, p. 10.
11. Kainth, G.S. and Balbir Singh, Pricing Efficiency in Foodgrains Marketing System - The Nature of Spatial and Temporal Price Relationship, *Agricultural Marketing,* Vol. XXII, No. 1, April, 1980, p. 6.
12. Gangawar, A.C., Shrinivas and Rai, K.N., Marketing Pattern of Gram in Selected Markets of Haryana, *Agricultural Marketing*, Vol. XXVI, No. 1, April, 1983, p. 14.
13. Hosamani, S.B., Hugar, L.B. and Hiremath, K.C., Structure and Movement of Market Arrivals and Price Behaviour of Cotton in Belgaum District, *Agricultural Marketing,* Vol. XXVIII, No. 2,. October November, 1985, p. 11.
14. Agrawal, N.L.,. Foodgrain Prices Analysis - An Interpretative Review of Studies in India, in *Agricultural Prices and Marketing in India,* Mittal Publications, Delhi, 1986, p. 10.
15. Kainth, G.S., Foodgrain Price Structure, in *Foodgrain Marketing System in India,*. Associated Publishing House, New Delhi, p. 46.
16. Raj Kumar Singh, Agricultural Price Behaviour During 1974-85, in. *Agricultural Price Policy in India,*. Printwell Publishers, Jaipur,. 1990, pp. 141-142.
17. Bhise, V.B., Behaviour of Market Arrival and Prices : A Study of Regulated Markets, in *Agricultural Marketing : Perspective and Issues,* Arihant Publishers, Jaipur, 1992, p. 67.

18. Rao, V.M., Market Arrivals and Price Movements, *Kurukshetra,* Vol. XLI, No. 6, March, 1993, p. 12.

19. Indrasena Reddy, Price Behaviour of Agricultural Commodities,. *Indian Journal of Marketing,* Vol. 23, No. 6-7, June-July, 1995, pp. 28-29.

20. Singh, H.S., Price Movement and Degree of Market Integration - A Study of Groundnut and. Jowar Market,. *Indian Society of Agricultural Economics Seminar Series V,* April, 1965.

21. Commission for Agricultural Costs and Prices, Price Behaviour of Kharif Crop Economy-An Overview,. *Report of the Agricultural Prices Commission on Price Policy for Crops Sown in 1984-85 Season,.* Department of. Agriculture and Cooperation, Ministry of Agriculture and Rural Development, New Delhi, 1984-85, p. 7.

22. Bhise, V.B.,. *op. cit.*, p. 65-66.

23. Katre, R.S., Thakare, R.P. and Borude, S.G., Trend and Variability in the Price and Arrivals of Rice (Paddy) in Selected Markets of Maharashtra State,. *Agricultural Situation in India,* Vol. XLVIII, No.2, May, 1993, p. 68.

23. Brahma Prakash and Sushila Srivastava, Analysis of Market Arrivals and Prices of Gram in Uttar Pradesh,. *Indian Journal of Agricultural Economics,* Vol. 48, No. 3, July-September, 1993.

24. Brahma Prakash and Sushila Srivastava, Pattern of Marketing Arrivals and Prices of Gram in Uttar Pradesh, *Indian Journal of Agricultural Marketing,* Vol. 8, No. 1, January-July, 1994, p. 114.

25. George, M.V. and Ranjit Singh, *op. cit.*, p. 3.

26. George, M.V. and Siraj Kunju, *op. cit.*, p. 12.

27. Adur, K.G., Price Behaviour of Agricultural Commodities, *Southern Economist,* Vol. No. XXIV, No. 2, 15 May, 1985.

29. Agarwal, N.L., *op. cit.*, p. 7.

30. Ibid., p. 8.

31. Ibid., p. 8.

32. Ibid., p. 9.

33. Kainth, G.S., *op. cit.*, p. 41-42.

34. Raj Kumar Singh, *op. cit.*, p. 136.

35. Katre, R.S., Thakare, R.P. and Borude, S.G., *op. cit.*, p. 67.

36. Babar, V.S. and Lohar, N.S., Trends in Arrivals and Prices of Jaggery in Sangli Regulated Market,. *Indian Journal of Agricultural Marketing,* Vol. 8, No. 1, January-July, 1994, p. 124.

37. Indrasena Reddy, *op. cit.*, p. 26.
38. George, M.V. and Ranjit Singh, *op. cit.*, p. 14.
39. Chauhan, K.K.S. and Verma, R.C., *op. cit.*, p. 17.
40. Verma, R.C. and Falor, L.N., A Study on Apple Price Behaviour and Market arrivals in Ajmer and Jaipur Markets, *Agricultural Marketing,* Vol. XX, No. 4, January, 1978, p. 12.
41. Jairth, M.S and Kansal, A.K., A Case Study of Price Behaviour of Himachal and Kashmir Apples in Delhi Market, *Agricultural Marketing,* Vol. XXIV, No. 2, July, 1981, p. 10.
42. Chatha, I.S. and Sidhu, D.S., Marketing of Kharif Pulses in Punjab, *Agricultural Marketing,* Vol. XXVI, January, 1984, p. 18.
43. Hosamani, S.B., Hugar, L.B. and Hiremath, K.C., *op. cit.*, p. 10.
44. Bhise, V.B., *op. cit.*, p. 69.
45. Rao, V.M., *op. cit.*, p. 13.
46. Brahm Prakash and Sushila Srivastava, *op. cit.*; p. 115.

3

SEASONAL VARIATION IN PRICES AND ARRIVALS OF AGRICULTURAL COMMODITIES

Agricultural production is of a seasonal nature. It is now the established fact that both prices and arrivals of agricultural commodities are subject to variation from one season to another within a year. A season can be either a day or a week or a month or a period of months which depends upon the duration required for producing the farm commodity and also on its perishability. Further, the period of a season differs from one agricultural commodity to another.

The seasonal (temporal or intra-year) variations are up and down swing in prices that occur with some regularity during the year. The seasonal variations resemble a cycle occupying a period of 12 months or less. The general pattern of seasonal variation in prices, lower prices during the post-harvest months and higher prices during pre-harvest or off-season months, is a normal feature for foodgrains and is repeated year after year due mainly to seasonality in suppiy and factors affecting the stocking decisions of the traders.

The seasonal nature of agricultural production, low capacity of the small and marginal farmers to hold stock and to obtain credit to postpone their sales, less storability of agricultural products are the factors which explain the variation in arrivals and prices of agricultural commodities from one season to another within a year.

One of the most important aspects of agricultural prices is the wide range of seasonal variations. The reason behind it is that the repayment of loans taken from money lenders, is made soonafter the harvesting of crops. Hence the farmers are forced to sell their produce immediately, after the harvest. Such sales are called distress sales. Thus the seasonal variation in prices is a common feature to almost all agricultural commodities. But the extent of seasonal variations differs from commodity to commodity and from market to market.

The study of seasonal variation in market arrivals and prices of agricultural commodities is important for providing guidance to farmers for marketing their farm products more efficiently.

In this chapter an attempt is made to study the seasonal variations in prices and arrivals of main foodgrains and oilseeds at regulated markets in Maratnwada. The foodgrains considered here for study include wheat, bajra, Kharif jowar, rabi jowar, tur and greengram. The oilseeds considered are sunflower, safflower and groundnut.

In order to study the intra-year variations the data on monthly arrivals and average prices of main foodgrains and oilseeds were obtained for chosen regulated marketing centres in Marathwada region. The seasonality of prices and arrivals of agricultural commodities is calculated in the form of range between maximum and minimum levels. Further, to study seasonal variation in arrivals and prices of foodgrains and oilseeds the measures like standard deviation and coefficient of variation are used.

Seasonal indices is a common feature to almost all agricultural commodities. But the extent of seasonal indices differ from commodity to commodity. The seasonal indices of arrivals and prices of selected commodities are calculated by Method of Simple Averages. This method of calculating seasonal indices comprises the ratio of a time series value to the average of a full year data around that value. Range, standard deviation, coefficient of variation and seasonal indices of monthly arrivals and average prices have been calculated separately for each commodity and market considered in the present study.

The discussion made in this chapter is divided into two sections. The first section analysis the seasonal variation in arrivals and prices of foodgrains and second section analysis the seasonal variation in arrivals and prices of oilseeds.

SECTION A : SEASONAL VARIATION IN ARRIVALS AND PRICES OF FOODGRAINS

This section deals with the seasonal variation in arrivals and prices of foodgrains, namely, wheat, bajra, kharif jowar, rabi jowar, tur, and greengram at the regulated markets in Marathwada.

Wheat

Wheat is the most important crop grown in India, as well as in Marathwada. The results obtained after analysing the data on monthly arrivals and average prices of wheat at regulated markets in Marathwada region are presented in table 1. They show that the range between the monthly maximum and minimum arrivals of wheat is the highest in Jalna (7679.46 quintals), followed by Latur (4522.07 quintals), Aurangabad (4107.40 quintals), Nanded (3419.80 quintals), Beed (2555.73 quintals), Lasur (2090.40 quintals), Manwat (1700.47 quintals), Manjlegaon (1171.20 quintals) and Naigaon (1032.87 quintals) markets. And the lowest range in arrivals is found in Osmanabad (436.67 quintals), followed by Kallam (809.40 quintals), Udgir (846.07 quintals), Parbhani (861.20 quintals) and Partur (963.07 quintals) markets. As far as monthly variations in prices is concerned it were the highest in Udgir (range was Rs. 91.87) market and the lowest in Latur (range was Rs. 43.80) market.

Both standard deviation and coefficient of variation in arrivals of wheat are found to be the highest in Jalna (2372.72) and Naigaon (104.61) markets and the lowest in Osmanabad (145.16) and Lasur (39.88) markets. Because the size of wheat arrivals is relatively higher in Jalna and Naigaon markets.

As far as the monthly variation in prices is concerned, the value of standard deviation and coefficient of variation are found to be highest in case of Partur market (i.e. 24.07 and 7.03 respectively) and the lowest in Latur (i.e. 12.57 and 3.58 respectively) market. Because in Partur market arrival of wheat is low as compared to Latur market and therefore the price fluctuations are different. Further, the data presented in table 1 reveals that the degree of variations in monthly arrivals is higher than that in monthly prices in case of wheat. It means that the arrivals of wheat have shown greater monthly variations than its prices.

Table 1

Seasonal Arrivals And Average Prices of Wheat at Regulated Markets in Marathwada During 1981 to 1995

(Arrivals in quintals)
(Price in Rs. per quintal)

MONTHS	MARKETS													
	AURANGABAD		LASUR		JALNA		PARTUR		PARBHANI		MANWAT		BEED	
	Arrivals	Prices	Arrivals	Prices	Arrivals	Prices	Arrivals	Prices	Arrivals	Prices	Arrivals	Prices	Arrivals	Prices
January	1418.33a	352.20	933.87a	346.80	2458.07	360.40	327.27	351.07	475.33	326.67	246.67	355.67	657.20a	358.40
February	1816.07	340.73	965.40	321.73	2839.73	338.73	329.67	331.93	622.60	337.13	281.20	329.87	798.47	374.33
March.	5525.73b	311.87	2202.73	310.20a	8002.53	316.20a	1044.07	310.73a	1172.07	320.13	965.60	308.67a	2632.73	326.60a
April.	5316.40	307.07	3024.27b	323.33	9634.53b	336.20	1130.67b	315.80	1172.47b	328.40	1840.80b	321.40	3212.93b	339.00
May	3510.33	303.80a	2316.20	318.60	5106.20	328.07	549.27	320.40	632.80	335.33	745.80	330.80	2012.07	338.87
June.	2840.74	308.07	1783.60	315.67	4039.60	325.67	488.87	325.27	496.07	333.87	492.60	325.07	1479.67	344.00
July	2083.93	325.60	1399.80	324.60	2920.40	331.93	373.73	338.07	381.33	352.67	355.33	341.33	1062.27	353.33
August	2362.73	327.33	1336.73	332.07	2893.13	335.93	395.87	333.47	400.87	347.87	298.93	342.40	999.07	358.80
September	2116.40	330.47	1178.27	332.60	1955.07a	344.60	205.07	387.40b	311.27a	315.87a	178.13	345.67	920.33	359.73
October	2014.73	329.13	1366.53	341.33	2282.93	356.20	222.20	352.93	371.73	330.60	217.07	346.87	932.93	365.83
November	1918.27	339.27	1193.47	359.53	2282.93	365.27	205.87	359.20	420.87	361.67	185.27	361.53	693.13	377.73
December	1571.53	368.20b	997.93	370.00b	2424.00	389.47b	167.60a	384.40	369.20	389.33b	140.33a	386.27b	678.40	395.40b
Average	2707.91	328.65	1558.73	333.04	3903.26	344.06	453.35	342.56	568.88	339.96	495.64	341.30	1339.93	357.63
Range between Maximum and Minimum	4107.40	64.40	2090.40	59.80	7679.46	73.27	963.07	76.67	861.20	73.46	1700.47	77.60	2555.73	68.80
S.D.	1464.63	18.66	621.39	17.45	2372.72	19.58	305.09	24.07	285.71	19.52	470.84	19.69	806.28	18.35
C.V.	54.09	5.68	39.88	5.24	60.79	5.69	67.30	7.03	50.22	5.74	94.99	5.77	60.17	5.13

contd.

MONTHS	*MARKETS*													
	MANJLEGAON		*LATUR*		*UDGIR*		*OSMANABAD*		*KALLAM*		*NANDED*		*NAIGAON*	
	Arrivals	*Prices*	*Arrivals*	*Prices*	*Arrivals*	*Prices*	*Arrivals*	*Prices*	*Arrivals*	*Prices*	*Arrivals*	*Prices*	*Arrivals*	*Prices*
January	484.07	355.00	1291.87	363.20	118.60[a]	324.33	120.27	364.53	172.53[a]	331.87	848.93	334.73	106.47	271.00
February	379.93	330.93	2188.47	343.87	302.87	345.60	215.73	345.27	260.07	327.47	689.87[a]	329.33	105.60	263.27
March.	909.40	309.87[a]	5633.80[b]	332.67[a]	964.67[b]	322.67	523.33	328.00[a]	797.40	314.00[a]	3821.80	293.40[a]	989.13	283.47
April.	1315.13[b]	334.47	5099.67	340.07	660.93	329.33	545.20[b]	351.40	981.93[b]	338.00	3288.53	302.60	1129.60[b]	292.27
May.	779.07	331.80	2659.93	345.33	339.80	333.67	365.60	350.00	660.67	339.73	1357.20	303.53	415.73	287.67
June.	489.80	331.00	1837.27	333.47	335.20	335.20	276.27	346.60	474.60	339.47	1119.80	306.93	177.13	272.67
July	304.33	353.53	1254.00	349.67	271.00	337.67	160.87	363.47	316.73	354.07	994.87	323.20	154.40	240.20[a]
August	276.40	342.13	1425.47	354.00	217.73	259.00[a]	172.27	363.53	248.33	345.33	840.53	330.20	167.60	278.60
September	200.20	336.13	1111.73[a]	352.33	120.13	327.53	143.80	366.27	204.13	352.67	4109.67[b]	328.00	192.53	293.87
October	203.73	356.87	1134.40	357.00	137.73	324.00	227.80	374.47	280.73	360.07	917.87	325.93	152.20	294.73
November	162.13	365.07	1478.07	365.33	175.13	350.87[b]	135.20	378.40	215.07	343.33	1416.07	337.53	205.80	298.33[b]
December	143.93[a]	388.07[b]	1459.33	376.47[b]	143.53	336.20	108.53[a]	386.27[b]	194.80	377.67[b]	1062.80	351.80[b]	96.73[a]	280.73
Average	470.68	344.57	2214.50	351.12	315.61	327.17	249.57	359.85	400.58	343.64	1705.66	322.27	324.41	279.73
Range between maximum and minimum	1171.20	78.20	4522.07	43.80	846.07	91.87	436.67	58.27	809.40	63.67	3419.80	58.40	1032.87	58.13
S.D.	344.15	19.51	1478.99	12.57	243.45	22.16	145.16	15.59	258.24	15.72	1202.94	16.40	339.37	15.72
C.V.	73.12	5.66	66.79	3.58	77.14	6.77	58.16	4.33	64.47	4.57	70.53	5.09	104.61	5.62

a = Minimum level.
b = Maximum level
S.D. = Standard deviation
C.V. = Coefficient of variation.
Source: The data on monthly arrivals and average prices are compiled from the office records of regulated markets.

Seasonal indices of arrivals and prices of wheat are presented in table 2. The seasonal indices of arrivals of wheat are found to be the highest in all the markets in Marathwada during March, April and May and the lowest indices in October, November, December and January. But the seasonal price indices of this commodity are found to be the highest in November, December and January and the lowest in March, April and May. From this it is clear that when the arrival indices of wheat are at higher level at that time price indices of wheat are at a lower level and vice versa. It means that increase in arrivals leads to reduction in prices and decrease in arrivals of wheat leads to increase in prices.

Bajra

Bajra is the most important cereal crop grown in Marathwada. The results presented in table 3 about bajra commodity reveal that the range between monthly maximum and minimum arrivals is the highest in Jalna (22426.40 quintals) market, followed by Latur (9521.14 quintals) market and the lowest range in arrivals of bajra is found in Kallam (507.13 quintal) market. It means that arrivals of bajra are subject to relatively greater variation in Jalna and Latur markets and lesser variation in Kallam market.

The range between the maximum and minimum prices of bajra is found to be the highest in Beed (Rs. 225.07) market and the lowest in Latur (Rs. 27.13) market. The data presented in table 3 also shows that in case of bajra the degree of monthly variation in prices and arrivals varied from market to market. The arrivals were relatively unstable in Jalna market and price was relatively unstable in Beed market.

The standard deviation and coefficient of variation in arrivals are found to be the highest in Jalna (7138.86) and Manwat (167.18) markets respectively but they are found to be the lowest in Kallam (170.0) and Beed (46.25) markets respectively. It means the arrivals of bajra have varied to a greater extent in Jalna and Manwat markets and to a smaller extent in Kallam and Beed markets during the period of this study. In Jalna market the size of bajra arrivals is very high because the traders from neighbouring small markets use to resale bajra in this market.

In case of monthly price variation both standard deviation and coefficient of variation are found to be the highest in Beed (i.e. 79.77

Table 2

Seasonal Indices of Arrivals and Prices of Wheat For Regulated Markets in Marathwada During 1981 to 1995

MONTHS	*MARKETS*													
	AURANGABAD		*LASUR*		*JALNA*		*PARTUR*		*PARBHANI*		*MANWAT*		*BEED*	
	Arrivals	*Prices*	*Arrivals*	*Prices*	*Arrivals*	*Prices*	*Arrivals*	*Prices*	*Arrivals*	*Prices*	*Arrivals*	*Prices*	*Arrivals*	*Prices*
January	52.38	107.17	59.93	104.13	62.97	104.75	72.19	102.49	83.55	96.09	49.77	104.21	49.05	100.22
February	67.07	103.68	61.95	96.60	72.75	98.45	72.72	96.90	109.44	99.17	56.73	96.65	59.59	104.67
March.	204.06	94.90	141.36	93.14	205.02	91.90	230.30	90.71	206.03	94.17	194.82	90.44	196.48	91.32
April.	196.33	93.44	194.08	97.08	246.83	97.72	249.41	92.19	206.10	96.60	371.40	94.17	239.78	94.79
May.	129.63	92.44	148.64	95.66	130.82	95.35	121.16	93.53	111.24	98.64	150.47	96.72	150.16	94.76
June.	104.90	93.74	114.46	94.78	103.49	94.66	107.84	94.95	87.20	98.21	99.39	95.25	110.43	96.19
July	76.96	99.07	89.83	97.47	74.82	96.88	82.44	98.69	67.03	103.74	71.69	100.01	79.28	98.80
August	87.25	99.60	85.78	99.71	74.12	97.64	87.32	97.35	70.47	102.33	60.31	100.32	74.56	100.33
September	78.16	100.56	75.62	99.87	50.09	100.16	45.23	113.09	54.72	92.91	35.94	101.28	68.68	100.59
October	74.40	100.15	87.70	102.49	58.49	103.53	49.01	103.03	65.34	97.25	43.80	101.63	69.63	102.15
November	70.84	103.23	76.59	107.95	58.49	106.17	45.41	104.86	73.98	106.39	37.38	105.93	51.73	105.62
December	58.03	112.04	64.04	111.10	62.10	113.20	36.97	112.22	64.90	114.52	28.31	113.18	50.63	110.56

contd.

MONTHS	MARKETS													
	MANJLEGAON		LATUR		UDGIR		OSMANABAD		KALLAM		NANDED		NAIGAON	
	Arrivals	Prices	Arrivals	Prices	Arrivals	Prices	Arrivals	Prices	Arrivals	Prices	Arrivals	Prices	Arrivals	Prices
January	102.85	103.03	58.34	103.44	37.58	99.13	48.19	101.30	43.07	96.57	49.77	103.87	32.82	96.88
February	80.72	96.04	98.82	97.94	95.96	105.63	86.44	95.95	64.92	95.29	40.45	102.19	32.55	94.11
March.	193.21	89.93	254.40	94.75	305.65	98.62	209.69	91.15	199.06	91.37	224.07	91.04	304.90	101.34
April.	279.41	97.07	230.29	96.85	209.41	100.66	218.45	97.65	245.13	98.36	192.80	93.90	348.20	104.48
May. .	165.52	96.29	120.11	98.35	107.66	101.99	146.49	97.26	164.93	98.86	79.57	94.19	128.15	102.84
June.	104.06	96.06	82.97	94.97	106.21	102.45	110.70	96.32	118.48	98.79	65.65	95.24	54.60	97.47
July	64.66	102.60	56.63	99.59	85.87	103.21	64.46	101.01	79.07	103.04	58.33	100.29	47.59	85.87
August	58.72	99.29	64.37	100.82	68.99	79.16	69.03	101.02	61.99	100.49	49.28	102.46	51.66	99.59
September	42.53	97.55	50.20	100.35	38.06	100.11	57.62	101.78	50.96	102.63	240.94	101.78	59.35	105.05
October	43.28	103.57	51.23	101.68	43.64	99.03	91.28	104.06	70.08	104.78	53.81	101.14	46.92	105.36
November	34.45	105.95	66.75	104.05	55.49	107.24	54.17	105.15	53.69	99.91	83.02	104.74	63.44	106.65
December	30.58	112.62	65.90	107.22	45.48	102.76	43.49	107.34	48.63	109.90	62.31	109.16	29.82	100.36

Source: Indices are based on data on monthly arrivals and average prices compiled from the office records of regulated. markets.

Table 3

Seasonal Arrivals and Average Prices of Bajra at Regulated Markets in Marathwada During 1981 to 1995

(Arrivals in quintals)
(Price in Rs. per quintal)

MONTHS	MARKETS											
	AURANGABAD		LASUR		JALNA		PARTUR		MANWAT		BEED	
	Arrivals	Prices	Arrivals	Prices	Arrivals	Prices	Arrivals	Prices	Arrivals	Prices	Arrivals	Prices
January	3219.87	207.87[a]	4689.87	244.40	7574.00	226.20	231.27	214.53[b]	117.87	221.27	7467.53	223.53
February	1823.53	208.00	1504.33	252.40	4515.33	217.20	64.20	189.40	35.47	202.33	7021.00	410.27
March	869.67	212.40	1128.47	250.20	2663.27	222.60	19.07	185.87	12.60	218.40	3914.67	426.40[b]
April	1012.13	209.53	655.40	256.07	2150.20	227.00	12.67	198.07	9.73	207.40	3086.20	375.07
May	703.40[a]	246.07	596.60[a]	235.27	1616.93	225.47	11.13	166.87	4.00[a]	143.00	2280.20[a]	306.80
June	1418.87	255.87[b]	686.13	239.00	1642.13	226.07	6.87[a]	169.80	7.00	143.40	2571.33	297.27
July	1338.67	242.87	846.33	244.33	1616.07[a]	235.33	17.33	154.27[a]	13.47	179.93	2371.47	309.87
August	2018.00	230.00	1124.93	239.13	1761.47	225.33	7.07	163.20	4.80	63.73[a]	3432.07	225.87
September	3383.60	215.47	2024.13	214.40[a]	10872.60	208.67	212.27	173.67	104.87	194.53	3108.27	203.13
October	5162.60	214.73	5282.20[b]	229.27	24042.47[b]	207.00[a]	2010.87[b]	180.93	1250.20[b]	197.87	5996.87	201.33[a]
November	5347.27	230.87	4503.27	252.40	17990.60	230.33	1502.20	193.67	917.40	217.67	7929.53[b]	207.00
December	6247.20[b]	236.80	4320.80	266.67[b]	11893.73	239.87[b]	518.07	203.80	374.00	226.07[b]	7791.20	219.60
Average	2712.07	225.87	2280.21	243.63	7361.57	224.26	384.42	182.84	237.62	184.63	4747.53	283.84
Range between maximum and minimum	5543.80	48.00	4685.60	52.27	22426.40	32.87	2004.00	60.26	1246.20	162.34	5649.33	225.07
S.D.	1923.41	16.03	1763.63	13.09	7138.86	9.19	638.94	17.22	397.24	45.16	2195.78	79.77
C.V.	70.92	7.10	77.35	5.37	96.97	4.10	166.21	9.42	167.18	24.46	46.25	28.10

contd.

MONTHS	MARKETS							
	MANJLEGAON		LATUR		UDGIR		KALLAM	
	Arrivals	Prices	Arrivals	Prices	Arrivals	Prices	Arrivals	Prices
January	2449.87	233.40[b]	6602.07	207.67	488.33	200.93	161.80	195.27
February	1247.73	222.93	4240.13	207.40	200.73	198.20	81.33	201.20
March	636.47	224.80	3263.07	219.67	107.67	203.33	34.60	197.00
April	645.27	230.87	2592.27	225.60[b]	118.07	204.67	23.60	197.07
May	418.13	218.87	1561.53	222.07	53.27	183.87	16.20	183.73
June	396.53	226.13	1824.73	218.53	55.60	216.80	18.80	169.80[a]
July	426.00	227.13	1659.60	223.00	70.80	189.33	9.00	207.13
August	364.27[a]	204.20	1385.33[a]	220.47	40.73[a]	157.13[a]	8.87[a]	177.40
September	889.47	185.93[a]	1577.40	198.47[a]	74.27	186.53	192.40	171.60
October	4065.27	193.07	4326.33	208.07	454.73	203.27	516.00[b]	200.80
November	5035.93[b]	203.40	8650.60	221.00	1037.53[b]	213.13	416.47	218.00
December	3522.00	215.60	10906.47[b]	223.13	762.40	225.20[b]	326.00	223.20[b]
Average	1674.75	215.53	4049.13	216.26	288.68	198.53	150.42	195.18
Range between maximum and minimum	4671.66	47.47	9521.14	27.13	996.80	68.07	507.13	53.40
S.D.	1593.00	14.75	2994.70	8.20	314.50	17.09	170.00	16.26
C.V.	95.12	6.84	73.96	3.79	108.94	8.61	113.01	8.33

a = Minimum level.
b = Maximum level
S.D. = Standard deviation
C.V. = Coefficient of variation.
Source: As indicated in table 1

and 28.10 respectively) and the lowest in Latur (i.e. 8.20 and 3.79 respectively) market. It means the prices of bajra showed maximum monthly variation in Beed market and least variation in Latur market during the period under study. Bajra is the main short duration foodgrain crop grown in Beed market area. The growers of bajra in this market area have to sell their surplus bajra output soonafter the crop harvest. This explains highest degree of seasonal variation in bajra prices in Beed market.

Seasonal indices of monthly arrivals and prices of bajra are presented in table 4. This table indicates that the seasonal indices of arrivals are the highest in the month of November, December and January and the lowest in the month of March, April, May and June. Seasonal price indices are found to be the highest in the month of December, January and February and lowest in the month of September, October and November in most of the regulated markets considered. But in some markets the lowest indices are appeared in different months i.e. Manwat (May and June), Aurangabad (January and February), Kallam (May and June) and Udgir (July and August). Higher indices of arrival in the months of November, December and January are found as this period is the harvest period of bajra crop. All the markets showed inverse relationship between arrivals and prices except Aurangabad, Manwat, Kallam and Udgir. These markets does not show inverse relationship because there is no effective demand for bajra in the months of February, May, June July and August.

The above discussion reveals that the seasonal indices of bajra seems to have reached a peak level and trough in the same months in all the markets. But seasonal price indices are found to be the highest in the same period in all the markets but the period of the lowest seasonal price indices of bajra is different in different markets.

Kharif Jowar

Kharif Jowar is the main cereal crop of the Marathwada region. The results presented in the table 5 indicate that the range between the monthly maximum and minimum arrivals of kharif jowar is the highest in Latur (28508.34 quintal) market and the lowest in Beed (1311.34 quintal) market. The range in monthly prices of jowar is found to be the highest in Lasur (Rs. 80.00) market and lowest in

Table 4

Seasonal Indices of Arrivals and Prices of Bajra for Regulated Markets in Marathwada During 1981 to 1995

MONTHS	*MARKETS*											
	AURANGABAD		*LASUR*		*JALNA*		*PARTUR*		*MANWAT*		*BEED*	
	Arrivals	*Prices*	*Arrivals*	*Prices*	*Arrivals*	*Prices*	*Arrivals*	*Prices*	*Arrivals*	*Prices*	*Arrivals*	*Prices*
January	118.72	92.03	205.68	100.52	102.89	100.87	60.16	117.33	49.60	119.84	157.29	78.75
February	67.24	92.09	65.97	103.60	61.34	96.85	16.70	103.59	14.93	109.58	147.89	144.54
March.	32.07	94.04	49.49	102.70	36.18	99.26	4.96	101.66	5.30	118.29	82.46	150.22
April.	37.32	92.76	28.74	105.11	29.21	101.22	3.30	108.33	4.09	112.33	65.01	132.14
May. .	25.94	108.94	26.16	96.57	21.96	100.54	2.90	91.27	1.68	77.45	48.03	108.09
June.	52.32	113.28	30.09	98.10	22.31	100.81	1.79	92.87	2.95	77.67	54.16	104.73
July	49.36	107.52	37.12	100.29	21.95	104.94	4.51	84.37	5.67	97.45	49.95	109.17
August	74.41	101.83	49.33	98.15	23.93	100.48	1.84	89.26	2.02	34.52	72.29	79.58
September	124.76	95.39	88.77	88.00	147.69	93.05	55.22	94.98	44.13	105.36	65.47	71.56
October	190.36	95.07	231.65	94.11	326.59	92.31	523.09	98.96	526.14	107.17	126.32	70.93
November	197.17	102.21	197.49	103.60	244.39	102.71	390.77	105.92	386.08	117.89	167.02	72.93
December	230.35	104.84	189.49	109.46	161.57	106.96	134.77	111.46	157.40	122.44	164.11	77.37

MONTHS	MARKETS							
	MANJLEGAON		*LATUR*		*UDGIR*		*KALLAM*	
	Arrivals	*Prices*	*Arrivals*	*Prices*	*Arrivals*	*Prices*	*Arrivals*	*Prices*
January	146.28	108.29	163.05	96.03	169.16	101.21	107.56	100.04
February	74.50	103.43	104.72	95.90	69.53	99.83	54.07	103.08
March.	38.00	104.30	80.59	101.58	37.30	102.42	23.00	100.93
April.	38.53	107.12	64.02	104.32	40.90	103.09	15.69	100.97
May. .	24.97	101.55	38.56	102.69	18.45	92.61	10.77	94.13
June.	23.68	104.92	45.06	101.05	19.26	109.20	12.50	87.00
July	25.44	105.38	40.99	103.12	24.53	95.36	5.98	106.12
August	21.75	94.74	34.21	101.95	14.11	79.15	5.90	90.89
September	53.11	86.27	38.96	91.78	25.73	93.95	127.91	87.92
October	242.74	89.58	106.85	96.21	157.52	102.39	343.03	102.88
November	300.70	94.37	213.64	102.19	359.41	107.35	276.87	111.69
December	210.30	100.03	269.35	103.18	264.10	113.43	216.72	114.35

Source: As indicated in table 2

Table 5
Seasonal Arrivals and Average Prices of Kharif Jowar. At Regulated Markets in Marathwada During 1981 to 1995

(Arrivals in quintals)
(Price in Rs. per quintal)

MONTHS	MARKETS											
	AURANGABAD		*LASUR*		*JALNA*		*PARTUR*		*PARBHANI*		*BEED*	
	Arrivals	*Prices*	*Arrivals*	*Prices*	*Arrivals*	*Prices*	*Arrivals*	*Prices*	*Arrivals*	*Prices*	*Arrivals*	*Prices*
January	2179.73	182.93	973.53	207.73	6018.87	195.67	519.80	192.87	1050.13	180.27	1224.13	201.73
February	1744.53	185.27	418.60	196.40	4672.47	196.87	201.27	190.47	1100.13	185.07	656.40	196.80
March.	1082.13	186.00	288.67	175.73	2755.60	192.53	63.67	179.07	541.27	161.93	345.53	202.20
April.	1261.00	208.67	261.87	190.93	2895.87	196.87	38.00	144.07	337.47	175.75	318.93	209.60
May	911.13	194.13	224.47[a]	176.00	2782.93	201.67	26.67	153.40	145.40	168.33	256.07	213.53
June	1478.93	195.67	275.73	178.40	2810.67	209.73	16.93[a]	172.60	392.67	177.33	214.80	202.07
July	1259.33	214.07[b]	490.27	193.73	2556.20	217.13[b]	23.07	159.00	91.87[a]	179.47	207.07	221.87
August	1346.47	202.73	433.07	150.53	2686.13	202.60	37.47	141.27[a]	153.20	139.67[a]	288.87	225.73[b]
September	775.93[a]	198.27	285.00	135.27[a]	1613.13[a]	206.20	19.07	196.13	162.80	195.87[b]	167.13[a]	207.53
October	6082.40	182.80[a]	2133.93	185.80	10987.20	179.33[a]	1534.07	177.60	1162.27	170.27	956.47	192.27[a]
November	6849.47[b]	188.20	3394.87[b]	215.27[b]	15857.33[b]	185.93	2620.60[b]	187.20	2781.60[b]	175.13	1478.47[b]	195.93
December	3401.60	185.47	1511.00	214.73	7308.93	203.07	895.73	200.73[b]	1465.67	182.53	1109.07	206.33
Average	2364.39	193.68	890.92	185.04	5245.44	198.97	499.70	174.53	782.04	174.30	601.91	206.30
Range between maximum and minimum	6073.54	31.27	3170.40	80.00	14244.20	37.80	2603.67	59.46	2689.73	56.20	1311.34	33.46
S.D.	1983.24	10.04	946.38	23.15	4097.28	9.80	782.43	19.70	755.59	13.31	447.18	9.72
C.V.	83.88	5.18	106.23	12.51	78.11	4.92	156.58	11.29	96.62	7.64	74.29	4.71

contd.

MONTHS	MARKETS									
	LATUR		UDGIR		OSMANABAD		KALLAM		NAIGAON	
	Arrivals	*Prices*	*Arrivals*	*Prices*	*Arrivals*	*Prices*	*Arrivals*	*Prices*	*Arrivals*	*Prices*
January	25826.67	191.40	5782.60	178.27	1503.20	190.87	2574.73	188.20	1016.53	170.47
February	16076.53	194.33	5040.13	175.,40	843.27	197.60	2617.40	184.67	804.20	156.60
March.	12976.27	191.20	4586.67	169.40	427.80	198.33	1831.73	194.20	816.67	170.47
April.	12874.27	199.07	4092.20	181.13	313.40	197.27	1210.07	200.47	882.20	168.33
May	8866.00	203.13	3551.60	184.73	278.00	197.87	858.27	206.00	555.40	172.47
June	9662.60	206.93	4245.87	190.67[b]	257.07	199.00	676.80	215.53	318.47[a]	167.60
July	7412.27	216.60[b]	4363.47	185.67	274.47	199.93	1175.73	219.33[b]	353.00	157.00
August	8413.67	212.87	4052.80	186.40	298.87	209.27	633.13	218.53	501.00	151.00[a]
September	6339.93[a]	200.80	3070.93[a]	175.33	209.47[a]	214.47[b]	472.73[a]	213.67	473.87	175.33[b]
October	12159.67	184.00[a]	3366.33	166.40[a]	884.47	177.67[a]	1975.80	181.27[a]	1223.60	158.07
November	31930.93	188.80	73.18.27	177.87	2352.60[b]	191.27	4585.93[b]	195.27	2104.47[b]	167.13
December	34848.27[b]	203.33	8611.73[b]	179.73	2148.13	195.40	4224.27	199.93	1923.13	171.80
Average	15615.59	199.37	4840.22	179.25	815.90	197.41	1903.05	201.42	914.38	165.52
Range between maximum and minimum	28508.34	32.60	5540.80	24.27	2143.13	36.80	4113.20	38.06	1786.00	24.33
S.D.	9368.32	9.45	1583.27	6.78	738.64	8.74	1317.22	12.71	557.28	7.45
C.V.	59.99	4.74	32.71	3.78	90.53	4.43	69.22	6.31	60.95	4.50

a = Minimum level.
b = Maximum level
S.D. = Standard deviation
C.V. = Coefficient of variation.
Source: As indicated in table 1.

Udgir (Rs. 24.27) market. It means that the arrivals of kharif jowar have been subject to greater variation in Lasur and lesser variation in Beed market during the study period. In case of kharif jowar, the degree of monthly variation in prices is found to be the highest in Lasur market and the lowest in Udgir market. From this it is clear that the arrivals and prices of Kharif jowar have varied in different degrees in different markets.

The standard deviation and coefficient of variation of arrivals of Kharif jowar are found to be the highest in Latur (9368.32) market and Partur (156.58) market respectively and the lowest in Beed (447.18) market and Udgir (32.71) market respectively. It means that month to month variations in arrivals of Kharif jowar have been relatively higher in Latur and Partur markets and relatively lower in Beed and Udgir markets.

As far as monthly variation in prices is concerned it is found that the value of standard deviation and coefficient of variation are the highest for Lasur (i.e. 23.15 and 12.51 respectively) market and the lowest for Udgir (6.78 and 3.78 respectively) market. It means the month to month variations in prices of Kharif jowar have been relatively higher in Lasur market and relatively lower in Udgir market. The degree of seasonal variation of prices of kharif jowar found in case of Lasur market may be due to low production and arrivals of this commodity in this market area.

The seasonal indices of arrivals and prices of Kharif jowar are shown in table 6. The indices of arrivals of this commodity are found to be the highest in all markets (except Aurangabad and Lasur) in the months of October, November and December and the lowest indices in the months of June, July, August and September. In case of Aurangabad and Lasur markets the arrivals are lower as compared to other markets due to low acreage under kharif jowar in these market areas.

The indices of prices are found to be higher in different markets in different months i.e. Aurangabad (April and May), Lasur (November and December), Jalna (June and July), Partur (December and January), Parbhani (January and February), Beed (July and August), Latur (July and August), Udgir (May and June), Osmanabad (August and September), Kallam (July and August) and Naigaon (December and January) and lower in Aurangabad (December and

Table 6

Seasonal Indices of Arrivals and Prices of Kharif Jowar for Regulated Markets in Marathwada During 1981 to 1995

MONTHS	MARKETS											
	AURANGABAD		LASUR		JALNA		PARTUR		PARBHANI		BEED	
	Arrivals	Prices	Arrivals	Prices	Arrivals	Prices	Arrivals	Prices	Arrivals	Prices	Arrivals	Prices
January	92.19	94.45	109.27	112.26	114.74	98.34	104.02	110.51	134.28	103.42	203.37	97.79
February	73.78	95.66	46.99	106.14	89.08	98.95	40.28	109.13	140.67	106.18	109.05	95.40
March.	45.77	96.03	32.40	94.97	52.53	96.76	12.74	102.60	69.21	92.90	57.41	98.01
April.	53.33	107.74	29.39	103.18	55.21	98.95	7.60	82.55	43.15	100.83	52.99	101.60
May.	38.54	100.23	25.20	95.11	53.05	101.36	5.34	87.89	18.59	96.57	42.54	103.51
June.	62.55	101.03	30.95	96.41	53.58	105.41	3.39	98.89	50.21	101.74	35.69	97.95
July	53.26	110.53	55.03	104.69	48.73	109.13	4.62	91.10	11.75	102.97	34.40	107.55
August	56.95	104.67	48.61	81.35	51.21	101.83	7.50	80.94	19.59	80.13	47.99	109.42
September	32.32	102.37	31.99	73.10	30.75	103.64	3.82	112.37	20.82	112.37	27.77	100.60
October	257.25	94.38	239.52	100.41	209.46	90.13	307.00	101.76	148.62	97.69	158.91	93.20
November	289.69	97.17	381.05	116.33	302.31	93.45	524.44	107.26	355.69	100.48	245.63	94.97
December	143.87	95.76	169.60	116.04	139.34	102.06	179.26	115.01	187.42	104.72	184.26	100.01

contd.

MONTHS	MARKETS									
	LATUR		UDGIR		OSMANABAD		KALLAM		NAIGAON	
	Arrivals	*Prices*	*Arrivals*	*Prices*	*Arrivals*	*Prices*	*Arrivals*	*Prices*	*Arrivals*	*Prices*
January	165.39	96.00	119.47	99.45	184.24	96.69	135.29	93.44	111.17	102.99
February	102.95	97.47	104.13	97.85	103.36	100.09	137.54	91.68	87.95	94.61
March	83.10	95.90	94.76	94.50	52.43	100.46	96.25	96.41	89.31	102.99
April.	82.44	99.85	84.55	101.05	38.41	99.93	63.59	99.53	96.48	101.70
May	56.78	101.89	73.38	103.06	34.07	100.23	45.10	102.27	60.74	104.20
June	61.88	103.79	87.72	106.37	31.51	100.80	35.56	107.00	34.83	101.26
July	47.47	108.64	90.15	103.58	33.64	101.28	61.78	108.89	38.61	94.85
August	53.88	106.77	83.73	103.99	36.63	106.01	33.27	108.49	54.79	91.23
September	40.60	100.72	63.45	97.81	25.67	108.64	24.84	106.08	51.82	105.93
October	77.87	92.29	69.55	92.83	108.40	90.00	103.82	89.99	133.82	95.50
November	204.48	94.70	151.20	99.23	288.35	96.89	240.98	96.95	230.15	100.97
December	223.16	101.99	177.92	100.27	263.28	98.98	221.97	99.26	210.32	103.79

Source: As indicated in table 2.

January), Lasur (August and September), Jalna (October and November), Partur (April and May), Parbhani(October and November), Beed (October and November), Latur (October and November), Udgir (September and October), Osmanabad (October and November), Kallam (October and November) and Naigaon (July and August) markets. It means that the months of highest and lowest indices of prices differed from market to market. This may be due to the use of different seed varieties of this crop in different market areas.

Rabi Jowar

Rabi jowar is an important crop of the Marathwada region. In Maharashtra state, rabi jowar is grown mainly in Marathwada region. The results presented in table 7 show that the range between monthly maximum and minimum arrivals is the highest in Jalna (13448.20 quintals) market and the lowest in Udgir (354.079 quintals) market. In the case of monthly prices of rabi jowar the range between the maximum and minimum prices is found to be the highest in Udgir (Rs. 141.60) market and the lowest in Nanded (Rs. 33.80) market.

The monthly variations in arrivals and prices of rabi jowar are measured by calculating the standard deviation and coefficient of variation in its arrivals and prices. The standard deviation in arrivals is found to be the highest in Jalna (4399.60) market whereas the coefficient of variation in arrivals is found to be the highest in Udgir (72.55) market. On the other hand, the lowest value of coefficient of variation is found for Parbhani (29.10) market. But the interesting thing to be noted here is that the standard deviation and coefficient of variation in monthly prices of rabi jowar are found to be the highest in Udgir (i.e. 38.38 and 12.54 respectively) market and the lowest in Nanded (8.70 and 4.11 respectively) market. From this it can be said that prices of rabi jowar have been subject to lesser variation in Nanded market during the period of this study. And the prices have varied to a larger extent in Udgir market.

The monthly variations in arrivals and prices of rabi jowar are also studied by calculating the seasonal indices in its arrival and prices. These seasonal indices are present in table 8. The indices of arrivals in all selected markets (except Nanded) are the highest in the months of March, April and May and the lowest in November,

Table 7

Seasonal Arrivals and Average Prices of Rabi Jowar. At Regulated Markets in Marathwada During 1981 to 1995

(Arrivals in quintals)
(Price in Rs. per quintal)

MONTHS	MARKETS													
	AURANGABAD		LASUR		JALNA		PARTUR		PARBHANI		MANWAT		BEED	
	Arrivals	Prices	Arrivals	Prices	Arrivals	Prices	Arrivals	Prices	Arrivals	Prices	Arrivals	Prices	Arrivals	Prices
January	1876.07	312.00	1780.73	303.80	3746.27[a]	307.80	567.13	291.73	533.53	335.60	456.13	285.33	2000.67[a]	303.93
February	1703.73	302.33	1600.27[a]	284.20[a]	4596.73	293.80[a]	469.13[a]	281.07	793.27	289.00	561.60	278.07[a]	2076.47	296.53
March.	3464.73	297.33[a]	2961.27	295.93	15627.20	295.27	1580.07	276.40	851.27	304.07	1951.73	286.73	5385.93	294.47[a]
April.	3961.87[b]	309.13	4143.93	308.33	17194.47[b]	304.13	2510.07[b]	289.87	1255.13[b]	275.07[a]	2983.00[b]	297.27	7067.40[b]	305.13
May	2639.80	305.13	4070.20	300.87	10728.13	297.20	1572.27	275.47[a]	764.20	313.33	1672.87	291.13	5326.00	296.53
Jun	2294.27	307.27	4513.87[b]	309.13	9324.33	298.67	1475.13	287.40	659.40	310.20	1455.93	283.13	4731.27	297.67
July	2017.60	308.53	4444.73	306.67	8538.40	313.07	1545.40	294.27	592.87	312.67	1233.80	304.33	4654.47	306.60
August	1844.13	314.20	4025.73	313.13	7892.13	297.40	1294.93	285.60	703.07	362.93[b]	1062.67	298.87	4116.47	304.00
September	1881.47	309.87	3990.40	313.40	4829.87	308.60	666.27	299.47	614.53	319.93	655.87	299.67	3557.20	299.07
October	1470.47	310.87	3233.13	337.47	4688.80	309.40	677.27	291.40	649.53	309.20	712.20	292.40	2846.27	314.60
November	1022.47[a]	331.07	2517.07	359.00	4150.27	333.47	583.33	300.93	482.60	321.33	958.00	322.33	2277.47	339.13
December	1230.53	348.27[b]	2217.60	364.93[b]	4182.00	345.33[b]	486.27	334.67[b]	470.60[a]	353.33	370.07[a]	337.40[b]	2111.13	340.47[b]
Average	2117.26	313.00	3291.58	316.41	7958.22	308.68	1118.94	292.36	697.50	317.22	1172.82	298.06	3845.90	308.18
Range between maximum and minimum	2939.40	50.94	2913.60	80.73	13448.20	51.53	2040.94	59.20	784.53	87.86	2612.93	59.33	5066.73	46.00
S.D.	977.95	13.15	1008.00	23.63	4399.60	15.16	612.85	14.90	202.97	23.54	723.80	16.26	1568.95	15.12
C.V.	46.19	4.20	30.62	7.47	55.28	4.91	54.77	5.10	29.10	7.42	61.71	5.46	40.80	4.90

contd.

MONTHS	MARKETS											
	MANJLEGAON		LATUR		UDGIR		OSMANABAD		KALLAM		NANDED	
	Arrivals	Prices	Arrivals	Prices	Arrivals	Prices	Arrivals	Prices	Arrivals	Prices	Arrivals	Prices
January	1658.67	284.07	1052.00	333.33	67.93	342.00	271.20[a]	304.87	187.60[a]	302.53	1853.53	208.73
February	1235.40[a]	278.20[a]	900.20[a]	317.80	50.13[a]	312.27	288.47	299.20	193.73	286.27[a]	1151.47[a]	206.13
March	3924.33	294.07	1697.93	317.33	229.07	304.87	505.53	290.53[a]	435.00	295.20	1314.60	208.47
April	6703.87[b]	305.40	3187.87[b]	328.47	404.20[b]	330.27	886.33[b]	314.20	951.27[b]	319.60	1582.93	214.47
May	4164.27	300.53	1892.87	334.33	128.20	295.93	660.00	304.47	748.40	316.67	1332.00	211.20
June	3422.40	300.47	2121.13	320.67	131.40	291.07	513.60	312.33	577.47	311.40	1730.33	214.00
July	3207.07	305.00	1630.33	324.80	127.93	259.87	406.33	326.20	477.47	312.53	2018.73	228.33[b]
August	2766.40	295.87	1418.13	313.13[a]	102.60	287.00	532.33	306.67	454.60	303.40	2166.60	216.27
September	2260.33	295.20	1753.13	324.87	78.80	243.27[a]	530.07	313.27	337.00	310.60	5906.07[b]	222.80
October	2114.47	300.53	1527.20	330.67	53.47	273.80	424.73	319.73	327.53	323.27	2027.13	194.53[a]
November	1884.67	334.93	1359.20	361.33	116.27	347.73	306.20	350.47	205.67	339.33	2987.80	200.47
December	1332.07	337.67[b]	1221.40	363.67[b]	76.00	384.87[b]	334.13	358.67[b]	217.40	348.33[b]	2452.60	213.73
Average	2889.50	302.66	1646.78	330.87	130.50	306.08	471.58	316.72	426.10	314.09	2210.32	211.59
Range between maximum and minimum	5468.47	59.47	2287.67	50.54	354.07	141.60	615.13	68.14	763.67	62.06	4754.60	33.80
S.D.	1481.22	16.87	571.26	15.47	94.68	38.38	169.29	19.19	228.55	16.66	1219.29	8.70
C.V.	51.26	5.57	34.69	4.68	72.55	12.54	35.90	6.06	53.64	5.30	55.16	4.11

a = Minimum level.
b = Maximum level
S.D. = Standard deviation
C.V. = Coefficient of variation.

Source: As indicated in table 1.

Table 8

Seasonal Indices of Arrivals and Prices of Rabi Jowar for Regulated Markets in Marathwada During 1981 to 1995

MONTHS	MARKETS													
	AURANGABAD		*LASUR*		*JALNA*		*PARTUR*		*PARBHANI*		*MANWAT*		*BEED*	
	Arrivals	*Prices*	*Arrivals*	*Prices*	*Arrivals*	*Prices*	*Arrivals*	*Prices*	*Arrivals*	*Prices*	*Arrivals*	*Prices*	*Arrivals*	*Prices*
January	88.61	99.68	54.10	96.02	47.07	99.72	50.68	99.79	76.49	105.79	38.89	95.73	52.02	98.62
February	80.47	96.59	48.62	89.82	57.76	95.18	41.93	96.14	113.73	91.10	47.88	93.29	53.99	96.22
March	163.64	94.99	89.97	93.53	196.37	95.66	141.21	94.54	122.05	95.85	166.41	96.20	140.04	95.55
April	187.12	98.76	125.89	97.45	216.06	98.53	224.33	99.15	179.95	86.71	254.34	99.74	183.76	99.01
May	124.68	97.49	123.65	95.09	134.81	96.28	140.51	94.22	109.56	98.77	142.64	97.68	138.49	96.22
June	108.36	98.17	137.13	97.70	117.17	96.76	131.83	98.30	94.54	97.79	124.14	94.99	123.02	96.59
July	95.29	98.57	135.03	96.92	107.29	101.42	138.11	100.65	85.00	98.57	105.20	102.11	121.02	99.49
August	87.10	100.38	122.30	98.96	99.17	96.35	115.73	97.69	100.80	114.41	90.61	100.27	107.04	98.64
September	88.86	99.00	121.23	99.05	60.69	99.97	59.54	102.43	88.10	100.85	55.92	100.54	92.49	97.04
October	69.45	99.32	98.22	106.66	58.92	100.23	60.53	99.67	93.12	97.47	60.73	98.10	74.01	102.08
November	48.29	105.77	76.47	113.46	52.15	108.03	52.133	102.93	69.19	101.30	81.68	108.14	59.22	110.04
December	58.12	111.27	67.37	115.34	52.55	111.87	43.46	114.47	67.47	111.38	31.55	113.20	54.89	110.48

contd.

MONTHS	MARKETS											
	MANJLEGAON		*LATUR*		*UDGIR*		*OSMANABAD*		*KALLAM*		*NANDED*	
	Arrivals	*Prices*	*Arrivals*	*Prices*	*Arrivals*	*Prices*	*Arrivals*	*Prices*	*Arrivals*	*Prices*	*Arrivals*	*Prices*
January	57.40	93.86	63.88	100.74	52.05	111.74	57.51	96.26	44.03	96.32	83.86	98.65
February	42.75	91.92	54.66	96.05	38.41	102.02	61.17	94.47	45.47	91.14	52.10	97.42
March	135.81	97.16	103.11	95.91	175.53	99.60	102.20	91.73	102.09	93.98	59.48	98.52
April	232.01	100.90	193.58	99.28	309.73	107.90	187.95	99.21	223.25	101.75	71.62	101.36
May	144.12	99.30	114.94	101.05	98.24	96.68	139.96	96.13	175.64	100.82	60.26	99.81
June	118.44	99.28	128.80	96.92	100.69	95.10	108.91	98.61	135.53	99.14	78.28	101.14
July	110.99	100.77	99.00	98.17	98.03	84.90	86.16	102.99	112.06	99.50	91.33	107.91
August	95.74	97.76	86.12	94.64	78.62	93.77	112.88	96.83	106.69	96.60	98.02	102.21
September	78.23	97.53	106.46	98.19	60.38	79.48	112.40	98.91	79.09	98.89	267.20	105.30
October	73.18	99.30	92.74	99.94	40.97	89.45	90.07	100.95	76.87	102.92	91.71	91.94
November	65.22	110.66	82.54	109.21	89.10	113.61	64.93	110.66	48.27	108.03	135.18	94.74
December	46.10	111.57	74.17	109.91	58.24	125.74	70.85	113.25	51.02	110.90	110.96	101.01

Source: As indicated in table 2.

December and January. In case of Nanded market the highest indices of arrivals are found in July and August and the lowest in February and March. In case of Nanded, arrivals in June and July are highest because June and July months are the harvest period for rabi jowar. Secondly in this district the production of rabi jowar is taken on irrigated lands. The price indices of rabi jowar are found to be the highest in the months of November, December, and January in all the markets (except Nanded) and the lowest in February, March, April and May (except in Udgir and Nanded markets).

Tur

Tur is the important pulse crop in Marathwada. In case of tur table 9 reveals that the range between the monthly maximum and minimum arrivals is the highest in Latur (32719.40 quintals) market and the lowest in Naigaon (520.30 quintals) market. But the price range is found to be the highest in Naigaon (Rs. 527.86) market and the lowest in Jalna (Rs. 114.33) market.

The standard derivation and coefficient of variation in monthly arrivals of tur are found to be the highest in Latur (10282.48) and Aurangabad (136.49) markets respectively and the lowest in Naigaon (196.06) and Nanded (80.30) markets respectively. It means that the arrivals of this commodity are subject to relatively greater variation in Latur and Aurangabad markets and relatively lesser variation in arrivals in Naigaon and Nanded markets. As far as monthly variation in prices of tur is concerned, both the standard deviation and coefficient of variation are the highest for Naigaon (i.e. 162.11 and 32.12 respectively) market and the lowest for Jalna (i.e. 38.28 and 5.29 respectively) market. This indicates that the prices of tur commodity have been subject to lesser variation in Jalna market and have varied to a greater extent in Naigaon market. But the monthly arrivals of tur have shown lesser stability in Latur and Aurangabad markets and greater stability in Nanded and Naigaon markets during the study period. The lowest price variation in Naigaon market may be due to low level of arrivals of this commodity.

Monthly indices of arrivals and prices of tur commodity are shown in table 10. It is clear from this table that the indices of arrivals are the highest in the months of December, January, February and March and the lowest in September, October and November in all

Table 9

Seasonal Arrivals and Average Prices of Tur At Regulated Markets in Marathwada During 1981 to 1995

(Arrivals in quintals)
(Price in Rs. per quintal)

MONTHS	MARKETS													
	AURANGABAD		LASUR		JALNA		PARTUR		PARBHANI		MANWAT		BEED	
	Arrivals	Prices	Arrivals	Prices	Arrivals	Prices	Arrivals	Prices	Arrivals	Prices	Arrivals	Prices	Arrivals	Prices
January	9727.93[b]	670.40	3645.20[b]	655.40	25605.20[b]	676.87	5985.67[b]	668.00	8018.40[b]	667.33	8806.07[b]	660.20	9697.67[b]	671.20
February	4013.33	649.73	2000.87	638.40	15428.13	671.00[a]	3226.13	643.47	5262.60	635.80[a]	5109.47	647.33	5729.93	669.87[a]
March.	1820.53	663.20	1162.60	658.00	8689.33	678.27	3420.60	665.47	3995.53	638.53	3122.13	648.27	3229.73	676.33
April.	1370.93	664.80	641.40	643.47	6571.73	684.47	1457.60	673.53	2341.20	645.47	2447.67	682.27	2095.07	684.93
May. .	596.20	639.80	350.67	669.07	3436.13	699.53	606.80	703.67	1746.47	656.13	1410.20	679.20	931.00	702.13
June.	559.40	694.20	303.27	735.47	2509.73	734.20	382.80	759.13	1195.73	706.80	1097.80	720.07	666.00	760.13
July	222.40	748.27	141.67	743.27	1743.40	767.53	168.87	807.53[b]	771.40	738.53	463.67	752.87	409.47	790.33[b]
August	132.93	660.67	64.33	714.80	1030.47	738.00	68.00	755.53	271.27	741.73	245.73	735.47	172.67	755.40
September	385.00	624.20[a]	30.47[a]	567.80[a]	742.00	756.87	16.73	576.73	226.93	680.33	104.53[a]	654.73	71.07	734.67
October	98.13[a]	648.60	125.27	582.93	585.73[a]	756.67	10.53[a]	544.47[a]	254.60	719.20	426.33	634.60	64.87[a]	717.27
November	538.47	738.27	462.07	680.80	815.73	738.00	43.67	569.20	211.40[a]	683.93	197.73	584.27[a]	299.27	731.87
December	9116.73	778.40[b]	2757.00	789.40[b]	2323.80	785.33[b]	2280.53	779.67	3314.40	755.47[b]	3769.73	757.33[b]	3963.73	768.67
Average	2381.83	681.71	973.74	673.23	5790.12	723.90	1472.33	678.87	2300.83	689.10	2266.76	679.72	2277.54	721.90
Range between maximum and minimum	9629.80	154.20	3614.73	221.60	25019.47	114.33	5975.14	263.06	7807.00	119.67	8701.54	173.06	9632.80	120.46
S.D.	3250.92	46.11	1145.51	61.91	7298.44	38.28	1821.16	82.54	2348.84	40.62	2510.74	50.26	2839.73	39.63
C.V.	136.49	6.76	117.64	9.20	126.05	5.29	123.69	12.16	102.09	5.90	110.76	7.39	124.68	5.49

contd.

MONTHS	MANJLEGAON		LATUR		UDGIR		OSMANABAD		KALLAM		NANDED		NAIGAON	
	Arrivals	Prices	Arrivals	Prices	Arrivals	Prices	Arrivals	Prices	Arrivals	Prices	Arrivals	Prices	Arrivals	Prices
January	7946 47[b]	646.67	34855.87[b]	690.40	14373.73[b]	663.07[a]	2386.07[a]	670.13	2553.47[b]	644.80	3820.67[b]	654.53	442.27	647.07
February	5566 73	639.80[a]	26763.73	686.33[a]	13908.53	670.13	1882.20	658.00	2047.47	638.80	3747.67	650.33[a]	523.80[b]	630.00
March	3444 93	664.87	20764.07	700.93	10607.47	686.73	1032.20	655.93[a]	1126.33	667.13	3783.27	656.53	490.13	628.00
April	2979 93	666.80	19087.87	707.73	7553.67	693.07	982.53	679.33	640.20	661.00	2248.73	671.27	311.93	635.13
May	1478 93	696.33	14259.40	729.00	6078.13	740.80	531.27	682.13	339.40	692. 80	1711.13	681.93	115.40	634.20
June	919 07	761.20	10474.27	768.00	3891.60	795.27	398.33	755.13	229.53	739.07	1190.60	726.60	68.33	655.93[b]
July	537 40	776.37[b]	6217.00	808.40[b]	1861.00	788.80	196.73	781.07[b]	111.67	754.20	890.33	759.53	32.80	495.33
August	182 80	744.13	3395.47	787.33	917.33	804.07[b]	116.47	727.73	39.67	685.60	531.53	741.13	13.07	440.87
September	76 73	731.73	2137.00	784.53	693.27	788.60	63.07[a]	770.27	16.80	616.53	609.93	746.60	19.93	448.00
October	57.27[a]	680.20	2136.47[a]	772.33	168.87[a]	744.93	82.80	712.27	10.60[a]	540.80[a]	293.80	766.93[b]	8.87	271.13
November	84 53	644.80	2166.20	776.33	173.73	756.33	117.13	713.27	56.67	710.13	235.40[a]	735.53	3.47[a]	128.07[a]
December	3572.40	661.33	10723.07	805.67	3770.67	802.47	1129.00	749.20	870.27	759.33[b]	936.73	738.40	87.20	443.33
Average	2237.27	692.85	12748.37	751.42	5333.17	744.52	743.65	712.87	670.17	675.85	1666.65	710.78	176.43	504.76
Range between maximum and minimum	7889.20	136.53	32719.40	122.07	14204.86	141.00	2386.07	125.14	2542.87	218.53	3585.27	116.60	520.33	527.86
S.D.	2431.57	46.25	10281.46	43.62	4995.55	51.39	731.49	42.25	813.64	59.90	1338.27	42.34	196.06	162.11
C.V.	108.69	6.67	80.65	5.80	93.67	6.90	98.37	5.93	121.41	8.86	80.30	5.96	111.12	32.12

a = Minimum level.
b = Maximum level
S.D. = Standard deviation
C.V. = Coefficient of variation.
Source: As indicated in table 1

Table 10

Seasonal Indices of Arrivals and Prices of Tur for Regulated Markets in Marathwada During 1981 to 1995

MONTHS	*MARKETS*													
	AURANGABAD		*LASUR*		*JALNA*		*PARTUR*		*PARBHANI*		*MANWAT*		*BEED*	
	Arrivals	*Prices*	*Arrivals*	*Prices*	*Arrivals*	*Prices*	*Arrivals*	*Prices*	*Arrivals*	*Prices*	*Arrivals*	*Prices*	*Arrivals*	*Prices*
January	408.42	98.34	374.35	97.35	442.22	93.50	406.54	98.40	348.50	96.84	388.49	97.13	425.80	92.98
February	168.50	95.31	205.48	94.83	266.46	92.69	219.12	94.79	228.73	92.26	225.41	95.24	251.58	92.79
March.	76.43	97.28	119.40	97.74	150.07	93.70	232.33	98.03	173.66	92.66	137.74	95.37	141.81	93.69
April	57.56	97.52	65.87	95.58	113.50	94.55	99.00	99.21	101.75	93.67	107.98	100.38	91.99	94.88
May. .	25.03	93.85	36.01	99.38	59.34	96.63	41.21	103.65	75.91	95.21	62.21	99.92	40.88	97.26
June.	23.49	101.83	31.15	109.24	43.35	101.42	26.00	111.82	51.97	102.57	48.43	105.94	29.24	105.30
July	9.34	109.76	14.55	110.40	30.11	106.03	11.47	118.95	33.53	107.17	20.46	110.76	17.98	109.48
August	5.58	96.91	6.61	106.17	17.80	101.95	4.62	111.29	11.79	107.64	10.84	108.20	7.58	104.64
September	16.16	91.56	3.13	84.34	12.81	104.56	1.14	84.95	9.86	98.73	4.61	96.32	3.12	101.77
October	4.12	95.14	12.86	86.59	10.12	104.53	0.72	80.20	11.07	104.37	18.81	93.36	2.85	99.36
November	22.61	108.30	47.45	101.12	14.09	101.95	2.97	83.85	9.19	99.25	8.72	85.96	13.14	101.38
December	382.76	114.18	283.14	117.25	40.13	108.49	154.89	114.85	144.05	109.63	166.31	111.42	174.04	106.48

contd.

MONTHS	MARKETS													
	MANJLEGAON		LATUR		UDGIR		OSMANABAD		KALLAM		NANDED		NAIGAON	
	Arrivals	Prices	Arrivals	Prices	Arrivals	Prices	Arrivals	Prices	Arrivals	Prices	Arrivals	Prices	Arrivals	Prices
January	355.19	93.33	273.41	91.88	269.52	89.06	320.86	94.00	381.02	95.41	229.24	92.09	250.67	128.19
February	248.82	92.34	209.94	91.34	260.79	90.01	253.91	92.30	305.51	94.52	224.86	91.50	296.88	124.81
March	153.98	95.96	162.88	93.28	198.90	92.24	138.80	92.01	168.07	98.71	227.00	92.37	277.80	124.42
April	132.20	96.24	149.73	94.19	141.64	93.09	132.12	95.29	95.53	97.80	134.93	94.44	176.80	125.83
May	66.10	100.50	111.85	97.02	113.97	99.50	71.44	95.69	50.64	102.51	102.67	95.94	65.41	125.65
June	41.08	109.87	82.16	102.21	72.97	106.82	53.36	105.93	34.25	109.35	71.44	102.23	38.73	129.95
July	24.02	112.05	48.77	107.58	34.89	105.95	26.45	109.57	16.66	111.59	53.42	106.86	18.59	98.13
August	3.17	107.40	26.63	104.78	17.20	108.00	15.66	102.08	5.92	101.44	31.89	104.27	7.41	87.34
September	3.43	105.61	16.76	104.41	13.00	105.92	8.48	108.05	2.51	91.22	36.60	105.04	11.30	88.76
October	2.56	98.17	16.76	102.78	3.17	100.05	11.13	99.92	1.58	80.02	17.63	107.90	5.03	53.72
November	3.78	93.06	16.99	103.32	3.26	101.59	15.75	100.06	8.46	105.07	14.12	103.48	1.97	25.37
December	159.68	95.45	84.11	107.22	70.70	107.78	151.82	105.10	129.86	112.35	56.20	103.89	49.42	87.83

Source: As indicated in table 2.

the selected markets in Marathwada. It means that the period of three months (i.e. December, January and February) is a harvest season and the three months, namely, September, October and November forms the off-season for this commodity. Price indices of tur are the highest in June, July and August in all the markets except in Lasur and Aurangabad markets and the lowest in January, February and March in nine markets. In the remaining markets its price indices are the lowest in September, October and November. Because harvest period of tur is different from market to market and therefore its price indices are to be different from market to market.

Greengram

Greengram is the important short-duration pulse crop grown in Marathwada. In case of greengram the range between the monthly maximum and minimum levels of arrivals is found to be the highest in Jalna (58198.66 quintals) market and the lowest in Kallam (313.87 quintals) market (see table 11). But the range in monthly prices of this commodity is found to be the highest in Aurangabad (Rs. 361.73) market and the lowest in case of Latur market (Rs. 70.46).

The standard derivation of monthly arrivals of greengram is found to be the highest in Jalna (16224.61) market and the lowest in Kallam (99.71) market. But the value of coefficient of variation in monthly arrivals of this commodity is found to be the highest in Aurangabad (190.34) market and the lowest in Nanded (101.56) market. From this it can be said that the arrivals of greengram have been subject to greater monthly variation in Jalna and Aurangabad markets. The variation in monthly arrivals of greengram have been of lesser degree in Kallam and Nanded markets during the period of this study. Both the standard deviation and coefficient of variation in monthly prices of greengram are found to be the highest in Aurangabad (i.e. 110.95 and 20.58 respectively) market and the lowest in Latur (i.e. 21.10 and 3.41 respectively) market. It means that the prices of greengram have been most unstable in Aurangabad market and least unstable in Latur market during the period of this study.

Seasonal indices of prices and arrivals of greengram are presented in table 12. The indices of monthly arrivals of greengram are found to be the highest in September and October and are lowest

Table 11

Seasonal Arrivals and Average Prices of Greengram at Regulated Markets in Marathwada During 1981 to 1995

(Arrivals in quintals)
(Price in Rs. per quintal)

MONTHS	MARKETS									
	AURANGABAD		JALNA		PARTUR		PARBHANI		MANWAT	
	Arrivals	Prices	Arrivals	Prices	Arrivals	Prices	Arrivals	Prices	Arrivals	Prices
January	28.40[a]	595.20	1898.80	589.60	35.60	544.80	1087.13	570.60	89.13	541.00
February	79.73	587.53	1191.00	615.40	10.20	492.00	2155.60	552.47	96.27	568.40
March	107.47	505.67	719.93	577.47	3.73	431.53	144.53	615.40	21.27	474.60
April	74.07	349.47	537.60	656.27[b]	9.73	342.47	101.00	574.47	58.53	527.73
May	56.07	272.67[a]	215.47[a]	598.60	1.97[a]	327.07[a]	28.53[a]	605.97	6.80[a]	323.87[a]
June	48.07	626.13	833.33	650.13	19.73	601.53	78.13	535.00	40.27	538.80
July	28.53	588.00	812.87	647.73	8.40	427.47	96.40	501.67[a]	32.27	565.40
August	2056.80	523.53	11859.67	563.53[a]	1619.13	537.47	2088.53	555.67	1745.80	545.67
September	5073.33[b]	535.07	58414.13[b]	596.27	6327.93[b]	574.87	21345.80[b]	599.73	9752.53[b]	573.27
October	883.87	624.07	23616.87	628.33	2367.13	632.87[b]	7729.87	604.47	3157.00	628.13[b]
November	222.67	634.40[b]	8964.93	647.47	734.73	614.87	3413.53	620.87	1406.27	606.93
December	77.60	626.53	3145.47	651.80	268.13	599.80	826.53	629.07[b]	555.40	617.20
Average	728.05	539.02	9350.84	618.55	950.53	510.58	3257.97	580.45	1413.38	542.58
Range between maximum and minimum	5044.93	361.73	58198.66	92.74	6325.96	305.80	21317.27	127.40	9745.73	304.26
S.D.	1385.78	110.95	16224.61	31.31	1780.43	101.46	5849.47	37.22	2684.28	77.38
C.V.	190.34	20.58	173.51	5.06	187.31	19.87	179.54	6.41	189.92	14.26

contd.

MONTHS	MARKETS							
	LATUR		UDGIR		KALLAM		NANDED	
	Arrivals	*Prices*	*Arrivals*	*Prices*	*Arrivals*	*Prices*	*Arrivals*	*Prices*
January	2239.60	577.67[a]	138.33	592.93	154.13	553.27	708.67	568.33
February	1079.20	585.93	87.80	610.60	97.60	596.20[b]	322.53	597.07
March	564.00	624.33	37.07	614.87	48.93	517.20	116.47	602.53
April	475.33	627.73	13.40[a]	533.33	16.80	370.47	115.40	592.87
May	232.93	622.00	26.33	622.33	5.13[a]	340.73[a]	62.53[a]	557.07[a]
June	420.67	640.80	76.73	629.73	7.53	522.67	74.13	678.80[b]
July.	412.73[a]	617.00	106.67	533.40	5.73	521.20	146.27	568.67
August	3327.47	593.67	1507.67	513.33[a]	66.00	534.33	221.20	583.93
September	14111.07[b]	614.00	5669.20[b]	604.93	260.07	575.33	1867.33[b]	591.93
October	6567.53	635.67	2123.07	637.40	116.40	493.60	1419.80	614.80
November	2779.67	631.67	844.20	636.27	23.13	547.33	1060.20	637.00
December	2507.27	648.13[b]	351.47	638.00[b]	319.00[b]	590.87	665.87	649.93
Average	2893.12	618.22	915.16	597.26	93.37	513.60	565.03	603.58
Range between maximum and minimum	13878.14	70.46	5655.80	124.67	313.87	255.47	1804.80	121.73
S.D.	3808.15	21.10	1574.61	43.06	99.71	76.74	573.87	34.59
C.V.	131.63	3.41	172.06	7.21	106.78	14.94	101.56	5.73

a = Minimum level.
b = Maximum level
S.D. = Standard deviation
C.V. = Coefficient of variation.
Source: As indicated in table 1

Table 12

Seasonal Indices of Arrivals and Prices of Greengram for Regulated Markets in Marathwada During 1981 to 1995

MONTHS	MARKETS									
	AURANGABAD		JALNA		PARTUR		PARBHANI		MANWAT	
	Arrivals	Prices	Arrivals	Prices	Arrivals	Prices	Arrivals	Prices	Arrivals	Prices
January	3.90	110.42	20.31	95.32	3.75	106.71	33.37	98.30	6.24	99.71
February	10.95	109.00	12.74	99.49	1.07	96.36	66.16	95.18	6.81	104.76
March	14.76	93.81	7.70	93.36	0.39	84.52	4.44	106.02	1.50	87.47
April	10.17	64.83	5.75	106.10	1.02	67.08	3.10	98.97	4.14	97.26
May	7.70	50.59	2.30	96.77	0.21	64.06	0.88	104.40	0.48	59.69
June	6.60	116.16	8.91	105.11	2.08	117.82	2.40	92.17	2.85	99.30
July	3.92	109.09	8.69	104.72	0.88	83.73	2.96	86.43	2.28	104.21
August	282.51	97.13	126.83	91.11	170.34	105.27	64.11	95.73	123.52	100.57
September	696.84	99.27	624.69	96.40	665.72	112.60	655.19	103.32	690.02	105.66
October	121.40	115.78	252.56	101.58	249.03	123.96	237.26	104.14	223.37	115.77
November	30.58	117.69	95.87	104.68	77.30	120.43	104.77	106.96	99.50	111.86
December	10.66	116.23	33.64	105.38	28.21	117.48	25.37	108.38	39.30	113.75

contd.

MONTHS	MARKETS							
	LATUR		UDGIR		KALLAM		NANDED	
	Arrivals	*Prices*	*Arrivals*	*Prices*	*Arrivals*	*Prices*	*Arrivals*	*Prices*
January	77.41	93.44	15.12	99.28	165.07	107.72	125.42	94.16
February	37.30	94.78	9.59	102.23	104.53	116.08	57.08	98.92
March	19.49	100.99	4.05	102.95	52.40	100.70	20.61	99.83
April	16.43	101.54	1.46	89.30	17.99	72.13	20.42	98.23
May	8.05	100.61	2.88	104.20	5.49	66.34	11.07	92.29
June	14.54	103.65	8.38	105.44	8.06	101.77	13.12	112.46
July	14.27	99.80	11.66	89.31	6.14	101.48	25.89	94.22
August	115.01	96.03	164.74	85.95	70.69	104.04	39.15	96.74
September	487.75	99.32	619.48	101.28	278.53	112.02	330.48	98.07
October	227.00	102.82	231.99	106.72	124.66	96.11	251.28	101.86
November	96.08	102.18	92.25	106.53	24.77	106.57	187.63	105.54
December	86.66	104.84	38.41	106.82	341.65	115.04	117.85	107.68

Source: As indicated in table 2.

in months of April, May and June in all the markets under consideration. It means that the highest indices of arrivals of greengram found in all the selected markets in September and October is due to the fact that the most part of total production of this commodity is sold soonafter its harvest.

The data shown in table 12 shows that monthly price indices of greengram are the highest in the months of October, November and December and the lowest in the months of April, May, June and July in all the selected markets except Latur. These indices are the lowest in the months of January and February. It means that arrivals are highest in the months of August, September and October because it is the harvest period of greengram. The price indices of this commodity are higher in October, November and December as it is the off- season of greengram.

SECTION B : SEASONAL VARIATION IN ARRIVALS AND PRICES OF OILSEEDS

This section deals with the monthly variation in arrivals and prices of oilseeds at regulated markets in Marathwada.

Safflower

Safflower is an important oilseed crop grown in rabi season on medium to black soils in the Marathwada region. The results presented in table 13 reveal that the range between the monthly maximum and minimum arrivals of Safflower is the highest in Manwat (6805.20 quintals) market and the lowest in Osmanabad (212.67 quintals) market. As far as monthly prices is concerned, the range between maximum and minimum is found to be the highest in Partur (Rs. 299.07) market and the lowest in Manjlegaon (Rs. 119.53) market.

The standard deviation in monthly arrivals is found to be the highest for Manwat (1959.79) market and the lowest for Osmanabad (60.97) market. But the coefficient of variation in monthly arrivals of Safflower is found to be the highest for Partur (153.43) market and the lowest for Jalna (64.12) market. It means that the variations in monthly arrivals have been of smaller degree in Osmanabad and Jalna markets and of greater degree in Manwat and Partur markets during the period of this study. But both standard deviation and

Table 13

Seasonal Arrivals and Average Prices of Safflower at Regulated Markets in Marathwada During 1981to 1995

(Arrivals in quintals)
(Price in Rs. per quintal)

MONTHS	MARKETS													
	AURANGABAD		LASUR		JALNA		PARTUR		PARBHANI		MANWAT		BEED	
	Arrivals	Prices	Arrivals	Prices	Arrivals	Prices	Arrivals	Prices	Arrivals	Prices	Arrivals	Prices	Arrivals	Prices
January	188.67[a]	661.07	300.93	630.20	1955.80	651.80	37.27	412.53[a]	126.87	546.80	51.87	466.87	71.47[a]	460.07[a]
February	1483.13	661.33	880.27	589.53	4546.13	589.93[a]	720.93	511.33	423.07	452.47	1318.40	544.07	998.40	525.60
March.	3228.40[b]	590.27[a]	2718.87[b]	610.47	7704.13[b]	593.93	2207.40[b]	593.53	2358.53[b]	580.60	6851.73[b]	584.20	3237.80[b]	589.87
April	1882.33	606.93	2146.33	620.00	5395.33	613.13	1175.47	628.20	2189.87	608.60	3715.80	624.67	2002.93	617.53
May. .	871.00	614.80	1258.87	582.47[a]	3090.40	637.73	347.40	646.00	946.53	634.00	1688.33	644.13	1237.13	631.93
June.	618.93	646.60	1001.13	689.07	2907.93	668.00	237.13	686.87	517.07	652.93	1034.47	675.33	968.47	657.13
July	328.27	684.80	667.93	719.80	2662.53	711.73	131.53	711.60[b]	317.40	687.07	463.87	697.67[b]	640.40	617.13
August	272.73	703.27	408.40	698.27	1792.93	699.67	43.67	514.87	168.53	555.93	214.73	601.47	455.73	634.20
September	220.27	688.80	291.67	684.47	1560.27	693.60	11.33	455.67	70.73[a]	407.33[a]	86.20	496.93	322.80	613.40
October	267.93	701.20	350.73	711.47	1526.13	705.93	39.00	627.07	132.47	540.40	139.93	625.07	370.40	694.60[b]
November	322.33	716.27	272.00	759.47[b]	1155.67[a]	729.07	35.33	668.60	124.93	688.80[b]	122.60	527.27	153.07	635.67
December	346.47	728.93[b]	260.93[a]	750.07	1307.00	739.33[b]	8.80[a]	633.13	84.27	661.67	46.53[a]	455.07[a]	82.27	625.60
Average	835.87	667.02	879.84	670.44	2967.02	669.49	416.27	590.78	621.69	584.72	1311.21	578.56	878.41	608.56
Range between maximum and minimum	3039.73	138.66	2457.94	177.00	6548.46	149.40	2198.60	299.07	2287.80	281.47	6805.20	242.60	3166.33	234.53
S.D.	882.73	43.05	770.56	58.98	1902.51	49.64	638.70	91.08	777.44	85.68	1959.79	76.87	894.22	58.77
C.V.	105.61	6.45	87.58	8.80	64.12	7.41	153.43	15.42	125.05	14.65	149.46	13.29	101.80	9.66

contd.

MONTHS	MARKETS							
	MANJLEGAON		LATUR		OSMANABAD		KALLAM	
	Arrivals	*Prices*	*Arrivals*	*Prices*	*Arrivals*	*Prices*	*Arrivals*	*Prices*
January	98.67	631.80	246.13[a]	628.60	4.40	463.87	78.40	591.33
February	757.13	580.87[a]	1281.40	557.47[a]	76.27	496.73	501.07	555.87[a]
March	4218.60[b]	629.80	2504.87[b]	561.47	215.20[b]	571.47	889.53[b]	567.20
April	2974.07	603.60	1800.00	590.53	121.27	587.40	411.73	580.80
May	1302.33	631.47	877.67	600.87	51.33	599.93	220.47	594.73
June	849.07	668.27	761.20	639.73	34.13	629.60[b]	107.53	644.80
July	445.47	700.40[b]	451.27	681.20	14.67	549.20	61.00	667.60
August	255.47	679.73	363.13	689.47	16.27	608.40	39.33	667.07
September	158.87	676.53	334.13	674.40	10.47	585.53	29.60	592.00
October	285.47	668.00	452.93	665.13	16.33	575.00	51.40	564.00
November	164.60	626.93	371.27	683.80	5.67	584.00	22.33	682.47[b]
December	78.00[a]	636.53	260.40	694.20[b]	2.53[a]	424.20[a]	12.53[a]	675.20
Average	965.65	644.49	808.70	638.91	47.38	556.28	202.08	615.24
Range between maximum and minimum	4140.60	119.53	2258.74	136.73	212.67	205.40	877.00	126.60
S.D.	1253.52	33.10	681.71	48.21	60.97	59.69	257.81	46.17
C.V.	129.81	5.14	84.30	7.55	128.69	10.73	127.58	7.50

a = Minimum level.
b = Maximum level
S.D. = Standard deviation
C.V. = Coefficient of variation.
Source: As indicated in table 1.

coefficient of variation in monthly prices of Safflower are found to be the highest in Partur (i.e. 91.08 and 15.42 respectively) market and the lowest in Manjlegaon (i.e. 33.10 and 5.14 respectively) market.

Seasonal indices of monthly arrivals and prices of Safflower are shown in table 14. The seasonal indices of arrivals of this commodity are found to be relatively higher in the months of March, April and May and relatively lower in November, December and January in all the markets except Aurangabad market. In case of Aurangabad market relatively lower indices of arrivals of this commodity are found in August and September. From this it can broadly be said that the indices of arrivals of safflower reach to a peak level in the months of March, April and May and to a minimum level in the months of November, December and January.

In case of price indices of safflower, it is found that they reach to a peak level in eight markets in the months of November and December. But in the remaining three markets the higher indices of its prices are found in July, August and lower indices in March and April. It means the period of highest and lowest indices in both arrivals and prices of this commodity differs from market to market because the time of harvest period is different from one market area to another.

Sunflower

Sunflower is the most important oilseed crop which is grown in all the seasons in the Marathwada region. In case of sunflower the range between the monthly maximum and minimum arrivals is found to be the highest in Latur (11260.82 quintals), followed by Beed (4472.82 quintals), Manjlegaon (3045.00 quintals) markets and the lowest in Osmanabad market (1069.40 quintals). In case of monthly prices, it is found to be the highest in Parbhani (Rs. 273.33), followed by Beed (Rs. 223.20) and Osmanabad (Rs. 184.60) markets and the lowest in Manjlegaon (Rs. 106.46), followed by Latur (Rs. 170.70) and Kallam (Rs. 182.66) markets.

The monthly variations in arrivals and prices of sunflower are measured by calculating the standard deviation and coefficient of variation in its arrivals and prices. The standard deviation in its arrivals is found to be the highest in Latur (3685.93) market and the

Table 14
Seasonal Indices of Arrivals and Prices of Safflower For Regulated Markets in Marathwada During 1981 to 1995

MONTHS	MARKETS													
	AURANGABAD		LASUR		JALNA		PARTUR		PARBHANI		MANWAT		BEED	
	Arrivals	Prices	Arrivals	Prices	Arrivals	Prices	Arrivals	Prices	Arrivals	Prices	Arrivals	Prices	Arrivals	Prices
January	22.57	99.11	34.20	94.00	65.92	97.36	8.95	69.83	20.41	93.52	3.96	80.69	8.14	75.60
February	177.44	99.15	100.05	87.93	153.22	88.12	173.19	86.55	68.05	77.38	100.55	94.04	113.66	86.37
March	386.23	88.49	309.02	91.06	259.66	88.71	530.28	100.46	379.37	99.30	522.55	100.97	368.60	96.93
April	225.19	90.99	243.95	92.48	181.84	91.58	282.38	106.33	352.25	104.088	283.39	107.97	228.02	101.47
May	104.20	92.17	143.08	86.88	104.16	95.26	83.46	109.35	152.25	108.43	128.76	111.33	140.84	103.84
June	74.05	96.94	113.79	102.78	98.01	99.78	56.97	116.26	83.17	111.67	78.89	116.73	110.25	107.98
July	39.27	102.67	75.92	107.36	89.74	106.31	31.60	120.45	51.05	117.50	35.38	120.59	72.90	101.41
August	32.63	105.43	46.42	104.15	60.43	104.51	10.49	87.15	27.11	95.08	16.38	103.96	51.88	104.21
September	26.35	103.26	33.15	102.09	52.59	103.60	2.72	77.13	11.38	69.66	6.57	85.89	36.75	100.80
October	32.05	105.12	39.86	106.12	51.44	105.44	9.37	106.14	21.31	92.42	10.67	108.04	42.17	114.14
November	38.56	107.38	30.91	113.28	38.95	108.90	8.49	113.17	20.10	117.80	9.35	91.13	17.43	104.45
December	41.45	109.28	29.66	111.88	44.05	110.43	2.11	107.17	13.56	113.16	3.55	78.66	9.37	102.80

contd.

MONTHS	*MARKETS*							
	MANJLEGAON		*LATUR*		*OSMANABAD*		*KALLAM*	
	Arrivals	*Prices*	*Arrivals*	*Prices*	*Arrivals*	*Prices*	*Arrivals*	*Prices*
January	10.22	98.03	30.44	98.39	9.29	83.39	38.80	96.11
February	78.41	90.13	158.45	87.25	160.98	89.30	247.96	90.35
March	436.87	97.72	309.74	87.88	454.22	102.73	440.19	92.19
April	307.99	93.65	222.58	92.43	255.96	105.59	203.75	94.40
May	134.87	97.98	108.53	94.05	108.34	107.85	109.10	96.67
June	87.93	103.69	94.13	100.13	72.04	113.18	53.21	104.77
July	46.13	108.67	55.80	106.62	30.96	98.73	30.19	108.51
August	26.46	105.47	44.90	107.91	34.34	109.37	19.46	108.42
September	16.45	104.97	41.32	105.56	22.10	105.26	14.65	96.22
October	29.56	103.65	56.0q	104.10	34.47	103.37	25.44	91.67
November	17.05	97.27	45.91	107.03	11.97	104.98	11.05	110.93
December	8.08	98.76	32.20	108.65	5.34	76.26	6.20	109.75

Source: As indicated in table 2.

Table 15

Seasonal Arrivals and Average Prices of Sunflower at Regulated Markets in Marathwada During 1981 to 1995

(Arrivals in quintals)
(Price in Rs. per quintal)

MONTHS	MARKETS											
	PARBHANI		BEED		MANJLEGAON		LATUR		OSMANABAD		KALLAM	
	Arrivals	Prices	Arrivals	Prices	Arrivals	Prices	Arrivals	Prices	Arrivals	Prices	Arrivals	Prices
January	737.47	743.27	1482.40	740.47	1374.00	698.00	1407.67	730.00	369.93	708.93	1440.47	696.13
February	341.67	677.73	1038.93	702.20	1644.20	659.73	10703.00	718.67[a]	485.33	700.73[a]	1023.07	680.93
March	322.53	598.20	1716.07	702.80	1610.67	703.33	6303.27	722.67	632.00	719.27	1247.87	697.53
April	719.00	694.87	2126.13	725.80	2484.53	712.13	9128.33	745.00	811.73	725.33	1151.07	721.33
May	877.13	714.67	1868.47	734.60	2450.60	737.87	7769.27	759.00	449.87	742.33	1055.27	743.40
June	423.07	677.00	1129.93	794.60	1639.40	727.00	4606.47	801.33	269.07	798.47	648.87	793.60
July	130.33	798.40[b]	509.27	823.73[b]	406.47	763.13[b]	2301.00	883.67	126.00	885.33[b]	248.27	855.93[b]
August	47.93	706.33	199.20	743.53	126.40[a]	735.67	1469.07	889.40[b]	97.53[a]	873.07	95.47[a]	844.33
September	21.20[a]	525.07[a]	180.13[a]	600.53[a]	133.07	664.73	1228.93[a]	817.00	169.93	774.67	232.40	772.47
October	113.13	532.87	2612.87	754.27	2290.93	689.67	5447.00	733.00	963.07	725.13	1492.07	673.27[a]
November	983.53	760.53	4653.00[b]	775.80	3171.40[b]	656.67[a]	8351.53	767.67	1166.93[b]	748.07	1744.60[b]	749.47
December	1679.93[b]	795.60	3275.80	799.13	2750.67	711.27	12489.75[b]	778.33	455.13	715.87	1734.00	770.80
Average	533.08	685.38	1732.68	741.46	1673.53	704.93	5933.77	778.81	499.63	759.77	1009.45	749.93
Range between maximum & minimum	1653.73	273.33	4472.87	223.20	3045.00	106.46	11260.82	170.73	1069.40	184.60	1649.13	182.66
S.D.	469.21	87.56	1259.09	55.87	979.56	31.97	3685.93	56.35	324.07	59.71	624.43	57.83
C.V.	88.02	12.78	72.67	7.54	58.53	4.54	62.12	7.24	64.86	7.86	61.86	7.71

a = Minimum level; b = Maximum level; S.D. = Standard deviation; C.V. = Coefficient of variation

Source: As indicated in table 1.

lowest in Osmanabad (324.07) market. The coefficient of variation in arrivals of this commodity is found to be the highest in Parbhani (88.02) market and the lowest in Manjlegaon (58.53) market. On the other hand, the standard deviation and coefficient of variation in prices of this commodity are found to be the highest in Parbhani (87.56 and 12.78 respectively) market and the lowest in Manjlegaon (31.97 and 4.54 respectively) market. It means the prices of sunflower have been most unstable in Parbhani market and least unstable in Manjlegaon market during the period of this study. In Latur and Parbhani markets the degree of variation in arrivals is higher than other markets because in these market areas the production of sunflower is higher as it is grown in all seasons on irrigated lands.

The data presented in table 16 shows that in case of sunflower the monthly variation in prices and arrivals varied from market to market. The arrival indices are higher in the months of November and December for all the selected markets and lowest in the months of June, July, August and September. As well as price indices are higher in the months of June, July and August for all the markets except Parbhani market. For Parbhani market relatively higher price indices of this commodity are found in November and December because it is a lean season for kharif sunflower.

Groundnut

Groundnut is the important oilseed crop grown in Marathwada. In case of groundnut the range between the monthly maximum and minimum levels of arrivals is found to be the highest in Nanded (13375.33 quintals) market and the lowest in Udgir (155.33 quintals) market. But the range of its monthly prices is found to be the highest in Jalna (Rs. 398.87) market and the lowest in Manjlegaon (Rs.100.80) market.

The standard deviation in monthly arrivals of this commodity is found to be the highest in Nanded (3720.43) market and the lowest in Udgir (44.32) market. But the coefficient of variation in monthly arrivals of groundnut is found to be the highest in Nanded (158.38) market and lowest in Latur (35.68) market. This indicates that the arrivals of groundnut have been subject to greater variation in Nanded market during the period under study. Because in Nanded market area, groundnut is taken in summer season on irrigated lands. So the

Table 16
Seasonal Indices of Arrivals and Prices of Sunflower for Regulated Markets in Marathwada During 1981 to 1995

MONTHS	MARKETS											
	PARBHANI		BEED		MANJLEGAON		LATUR		OSMANABAD		KALLAM	
	Arrivals	Prices	Arrivals	Prices	Arrivals	Prices	Arrivals	Prices	Arrivals	Prices	Arrivals	Prices
January	138.34	108.45	85.56	99.87	82.10	99.02	23.72	93.73	74.04	93.31	142.70	92.83
February	64.09	98.88	59.96	94.71	98.25	93.59	180.37	92.28	97.14	92.23.	101.35	90.80
March.	60.50	87.28	99.04	94.79	96.24	99.77	106.23	92.79	126.49	94.67	123.62	93.01
April	134.88	101.38	122.71	97.89	148.46	101.02	153.84	95.66	162.47	95.47	114.03	96.19
May	164.54	104.27	107.84	99.08	146.43	104.67	130.93	97.46	90.04	97.70	104.54	99.13
June	79.36	98.78	65.21	107.17	97.96	103.13	77.63	102.89	53.85	105.09	64.28	105.82
July	24.45	116.49	29.39	111.10	24.29	108.26	38.78	113.46	25.22	116.53	24.59	114.13
August	8.99	103.06	11.50	100.28	7.55	104.36	24.76	114.20	19.52	114.91	9.46	112.59
September	3.98	76.61	10.40	80.99	7.95	94.30	20.71	104.90	33.81	101.96	23.02	103.01
October	21.22	77.75	150.80	101.73	136.89	97.83	91.80	94.12	192.76	95.44	147.81	89.78
November	184.50	110.96	268.54	104.63	189.50	93.15	140.75	98.57	233.56.	98.46	172.83	99.94
December	315.14	116.08	189.06	107.78	164.36	100.90	210.49	99.94	91.09	94.22	171.78	102.78

Source: As indicated in table 2.

Table 17

Seasonal Arrivals and Average Prices of Groundnut at Regulated Markets in Marathwada During 1981 to 1995

(Arrivals in quintals)
(Price in Rs. per quintal)

MONTHS	MARKETS									
	JALNA		PARBHANI		MANWAT		BEED		MANJLEGAON	
	Arrivals	Prices	Arrivals	Prices	Arrivals	Prices	Arrivals	Prices	Arrivals	Prices
January	207.27	506.47	74.07	660.33	83.13	702.60	2189.33	736.73	1414.13	720.40
February	48.07	558.13	50.13	527.13	40.60	512.47	995.13	744.67	545.53	718.00
March	10.00[a]	332.13	20.93	476.80	14.53[a]	433.13[a]	629.47	742.20	274.00	709.67
April	10.80	331.33[a]	13.87[a]	430.80[a]	25.13	573.00	497.40	755.20	258.40[a]	746.13
May	50.60	596.00	1286.00	644.13	508.53	589.33	289.53[a]	721.47[a]	328.33	689.47
June	240.80	695.93	2767.81[b]	733.27	1523.00[b]	717.00	487.13	751.13	1073.07	727.47
July	435.20	727.27	1962.07	816.47[b]	601.53	783.47[b]	448.60	826.00[b]	1832.87	730.00
August	548.33	705.53	401.60	697.13	692.07	736.20	656.20	758.13	3257.47	715.07
September	676.67	652.67	221.07	578.67	1213.13	697.53	2127.47	733.87	3309.87	711.00
October	889.27	637.93	142.87	581.87	1419.60	711.53	3057.60	737.80	3992.83	705.13
November	1128.53[b]	707.20	162.67	732.07	1324.73	709.27	3612.00[b]	721.73	4249.33[b]	651.20[a]
December	685.80	730.20[b]	114.73	734.27	469.87	648.67	3377.27	761.93	3080.87	752.00[b]
Average	410.95	600.190	601.49	634.41	659.65	651.18	1530.59	749.24	1968.06	714.63
Range between maximum and minimum	1118.53	398.87	2753.94	385.67	1508.47	350.34	3322.47	104.53	3990.93	100.80
S.D.	360.81	136.02	870.90	112.10	552.97	99.00	1210.89	26.29	1461.63	25.20
C.V.	87.80	22.64	144.79	17.67	83.83	15.20	79.11	3.51	74.27	3.53

contd.

MONTHS	MARKETS									
	LATUR		UDGIR		OSMANABAD		KALLAM		NANDED	
	Arrivals	Prices	Arrivals	Prices	Arrivals	Prices	Arrivals	Prices	Arrivals	Prices
January	4616.60	723.73	65.27	699.40	247.73	704.80	160.80	699.20	570.53	705.60
February	4659.53	758.00	56.60	563.33	142.73	735.53	98.33	737.00	177.60	532.47
March	2450.20	778.13	22.40	479.00[a]	98.27[a]	766.73	157.40	763.60	110.00	516.60[a]
April	2932.00	788.73	25.07	629.07	107.53	782.67[b]	71.20[a]	793.33[b]	80.87[a]	620.40
May	4035.93	795.00	77.73	652.40	101.93	759.87	88.53	739.00	4354.53	751.33
June	5780.00[b]	800.33	88.13	757.67[b]	172.47	742.27	206.60	698.73	13456.20[b]	779.00
July	4526.87	833.33[b]	22.40[a]	661.93	134.93	715.33	353.67	691.47	5328.87	849.67
August	3498.60	799.07	37.00	667.80	278.13	708.47	583.00[b]	663.33[a]	1132.87	874.53[b]
September	3183.07	767.87	58.27	563.33	435.53[b]	636.93[a]	354.07	672.07	795.13	780.33
October	2727.00	766.40	95.47	713.33	303.13	683.53	395.20	703.40	685.53	797.33
November	3522.07	710.67[a]	124.40	707.47	409.20	719.27	308.73	710.73	809.20	767.73
December	666.33[a]	737.33	177.73[b]	710.27	400.27	728.53	274.00	773.00	687.67	767.73
Average	3549.85	771.55	70.87	650.42	235.99	723.66	254.29	720.41	2349.08	728.66
Range between maximum and minimum	5113.67	122.66	155.33	278.67	337.26	145.74	511.80	130.00	13375.33	357.93
S.D.	1266.70	33.76	44.32	76.46	122.34	37.60	146.29	39.06	3720.43	110.01
C.V.	35.68	4.38	62.54	11.75	51.84	5.20	57.53	5.42	158.38	15.10

a = Minimum level.
b = Maximum level
S.D. = Standard deviation
C.V. = Coefficient of variation.
Source: As indicated in table 1.

production of groundnut in this area depends largely on the availability of irrigation water. It is also found that the variations in monthly arrivals have been of smaller degree in Latur and Udgir markets. Both standard deviation and coefficient of variation in monthly prices of groundnut are found to be the highest in Jalna (i.e. 136.02 and 22.45 respectively) market and the lowest in Manjlegaon (25.20) and Beed (3.51) markets respectively. In case of Jalna market the degree of variation in monthly price is higher because the production of groundnut is lower in this market area. The data also indicates that both arrivals and prices of groundnut have varied in different degrees in different markets during the period of this study. Seasonal indices of arrivals and prices of groundnut in different markets are shown in table 18. This table shows that the highest arrival indices of groundnut are found for two seasons viz. November and December and June and July in all the markets. But the lowest indices of arrival of this commodity are found in March-April in all the markets except Latur market where it were the lowest in October and November. Price indices of groundnut are relatively higher in the months of June, July and August for some markets and during November and December for other markets. For Osmanabad market, the relatively higher price indices are seen in the months of March and April as it is the off-season for groundnut in this market area. From the above it is clear that the periods of the highest and lowest price indices of groundnut are different for different markets.

This study has indicated that the seasonal variation in arrivals of wheat were higher in Jalna and Naigaon markets and lower in Osmanabad and Lasur markets. In case of seasonal indices of arrivals of wheat it were relatively higher in the months of March, April and May and relatively lower in the months of October, November and December. This period must be the peak period for wheat commodity. The prices of wheat showed a higher degree of variation in Partur and lower in Latur market. The seasonal price indices of wheat were relatively higher in the months of November, December and January and relatively lower in the months of March, April and May.

In case of bajra the study indicated that the seasonal variation in arrivals were the highest in Jalna and Manwat markets and the lowest in Kallam and Beed markets. The seasonal indices of arrivals of bajra were relatively higher in the months of November, December

Table 18

Seasonal Indices of Arrivals and Prices of Groundnut for Regulated Markets in Marathwada During 1981 to 1995

MONTHS	*MARKETS*									
	JALNA		*PARBHANI*		*MANWAT*		*BEED*		*MANJLEGAON*	
	Arrivals	*Prices*	*Arrivals*	*Prices*	*Arrivals*	*Prices*	*Arrivals*	*Prices*	*Arrivals*	*Prices*
January	50.44	84.29	12.31	104.09	12.60	107.90	143.04	98.33	71.85	100.81
February	11.70	97 87	8.33	83.09	6.15	78.70	65.02	99.39	27.72	100.47
March	2.43	55.27	3.48	75.16	2.20	66.51	41.13	99.06	13.92	99.31
April	2.63	55.14	2.31	67.91	3.81	87.99	32.50	100.80	13.13	104.41
May	12.31	99.18	213.80	101.53	77.09	90.50	18.92	96.29	16.68	96.48
June	58.60	115.81	460.16	115.58	230.88	110.11	31.83	100.25	54.52	101.80
July	105.90	121.03	326.20	128.70	91.19	120.31	29.31	110.25	93.43	102.15
August	133.43	117.41	66.77	109.89	104.91	113.06	42.87	101.19	165.52	100.06
September	164.66	108.62	36.75	91.21	183.90	107.12	139.00	97.95	168.18	99.49
October	216.40	106.16	23.75	91.72	215.20	109.27	199.77	98.47	202.88	98.67
November	274.62	117.69	27.04	115.39	200.82	108.92	235.99	96.33	215.91	91.12
December	166.88	121.52	19.07	115.74	71.23	99.61	220.65	101.69	106.56	105.23

contd.

MONTHS	*MARKETS*									
	LATUR		*UDGIR*		*OSMANABAD*		*KALLAM*		*NANDED*	
	Arrivals	*Prices*	*Arrivals*	*Prices*	*Arrivals*	*Prices*	*Arrivals*	*Prices*	*Arrivals*	*Prices*
January	130.05	93.80	92.09	107.53	104.98	97.39	63.23	97.06	24.29	96.84
February	131.26	98.24	79.86	86.61	60.48	101.64	38.67	102.30	7.56	73.08
March	69.02	100.85	31.61	73.65	41.64	105.95	61.90	106.00	4.68	70.90
April	82.60	102.23	35.37	96.72	45.57	108.15	28.00	110.12	3.44	85.14
May	113.69	103.04	109.68	100.30	43.19	105.00	34.81	102.58	185.37	103.11
June	162.82	103.73	124.35	116.49	73.08	102.57	81.24	96.99	572.83	106.,91
July	127.52	108.01	31.61	101.77	57.18	98.85	139.08	95.98	226.85	116.61
August	98.56	103.57	52.21	102.67	117.86	97.90	229.26	92.08	48.23	120.02
September	89.67	99.52	82.22	86.61	184.56	88.01	139.24	93.29	33.85	107.09
October	76.82	99.33	134.71	109.67	128.45	94.45	155.41	97.64	29.18	109.42
November	99.22	92.11	175.53	108.77	173.40	99.39	121.41	98.66	34.45	105.36
December	18.77	95.56	250.77	109.20	169.61	100.67	107.75	107.30	29.27	105.53

Source: As indicated in table 2.

and January and relatively lower in the months of March, April and May. But the seasonal variation in its prices was the highest in Beed market and lowest in Latur market. Price indices of this commodity were relatively higher in the months of December, January, February and relatively lower in the months of September, October and November in case of most of the markets.

Kharif jowar arrivals have shown the highest degree of seasonal variation in Latur and Partur markets and the lowest variation in Beed and Udgir markets. Seasonal indices of arrivals of this commodity were comparatively higher in the months of October, November and December and lower in the months of June, July, August and September. Seasonal prices of kharif jowar have indicated larger degree of variation in Latur market and smaller degree of variation in Udgir market.

This study has indicated that the arrivals of rabi jowar have shown a higher degree of variation in Udgir market and lower degree of variation in Parbhani market. This is indicated by the value of coefficient of variation in arrivals for these markets. The seasonal indices of arrivals of rabi jowar were comparatively higher in the months of March, April and May in all the markets except Nanded. It were relatively lower in the months of November, December and January in all the markets. Price of jowar has varied to a greater extent in Udgir market and to a smaller extent in Nanded market. Seasonal price indices of this commodity are found to have reached to a higher level in the months of November, December and January in all the markets except Nanded. It were relatively lower in the months of February, March, April and May in all the markets except Udgir and Nanded markets.

Arrivals of tur have been subject to larger seasonal variation in Lasur and Aurangabad markets and smaller variation in Naigaon and Nanded markets. Seasonal indices of arrivals were relatively higher in the months of December, January and February and relatively lower in September, October and November in all the markets during the study period. But the seasonal variation in prices of this commodity was the highest in Naigaon and lowest in Jalna market. The seasonal price indices of this commodity were relatively higher in the months of June, July and August and relatively lower in January, February and March.

In case of greengram the seasonal variation of arrivals was the highest in Jalna and Aurangabad markets and lowest in Nanded and Kallam markets. The seasonal indices of arrivals of greengram were relatively higher in the months of September and October and relatively lower in the months of April, May and June in all the markets during the study period. The seasonal variation of its prices was the highest in Aurangabad market and lowest in Latur market. Seasonal price indices were relatively higher in the months of October, November and December and relatively lower in the months of April, May, June and July.

In case of safflower seasonal variation in arrivals was the highest in case of Manwat and Partur markets and the lowest in Osmanabad and Jalna markets. Seasonal indices of arrivals of safflower were relatively higher in the months of March, April and May and relatively lower in the months of November, December and January in all the markets except Aurangabad. But the degree of variation in its seasonal prices was the highest in Partur and lowest in Manjlegaon market. The periods of relatively higher and lower seasonal price indices were different for different markets.

In case of sunflower the seasonal variation of arrivals was the highest in Latur and Parbhani markets and the lowest in Manjlegaon market. Seasonal indices of arrivals of sunflower were relatively higher in the months of November and December and relatively lower in the months of June, July, August and September. On the other hand, the degree of seasonal price variation of this commodity was the highest in Parbhani and lowest in Manjlegaon market. The price indices were relatively higher in the months of June, July and August, and relatively lower in the months of November and December.

In case of groundnut seasonal variation of arrivals was the highest in Nanded and lowest in Udgir and Latur markets. But the variation in its seasonal prices was the highest in Jalna and lowest in Manjlegaon market. The periods of relatively higher and lower indices of both arrivals and prices of groundnut are found to be different for different markets.

4

ANNUAL VARIATION IN PRICES AND ARRIVALS OF AGRICULTURAL COMMODITIES

Agricultural prices are subject to wide fluctuations under certain conditions. They vary from year to year and also from season to season within a year. Annual variation in agricultural prices is due to several factors such as the changes in weather condition, size of the market, etc. Farmers do not have a full control over the output and yields of the crops. Because agricultural production is a biological process and depends upon the favourableness of the natural factors such as rainfall and climate. The fluctuations in crop yields, since they are unexpected and so cannot be counter balanced by varying acreages, inevitably affect prices. It means that variation in natural factors gives rise to variation in prices of farm products viz., variation in output and market supply (market arrival) of farm products. Acharya and Agarwal say that price variations are observed for agricultural commodities mainly because of the annual nature of agricultural production. The area under and yield of crops fluctuate from year to year, apart from other factors due to changes in weather. The components of weather which affect production of crops are commencement of rainfall, variation in minimum and maximum temperatures during various stages of crop growth, relative humidity, extent of sunshine, etc. The annual variation in prices also depends on size of the market. The annual fluctuations in prices are naturally less in the case of crop for which a world market is developed than the crops for which market is smaller. The magnitude of annual variation in agricultural prices no doubt depends upon total supply of farm products. However, local supplies often have a greater proportionate influence. Annual price variations result in fluctuations in farmers' incomes. If the demand for farm products is inelastic, the annual variations in yields cause prices to fluctuate. It means that when demand for farm products is inelastic, the producers incomes will vary from year to year with crop yields. The study of annual variations in market arrivals and prices of agricultural commodities is important for providing guidance to farmers for marketing their farm products more efficiently.

In this chapter an attempt is made to study the annual variations in prices and arrivals of main foodgrains and oilseeds at regulated markets in Marathwada. The foodgrains considered here for study include wheat, bajra, kharif jowar, rabi jowar, tur and greengram. The oilseeds include groundnut, safflower and sunflower. The study of annual variations in arrivals and prices of above mentioned foodgrains and oilseeds is made for chosen regulated marketing centres in Marathwada.

The annual variations in prices and arrivals of each selected commodity are measured through a range between the maximum and minimum levels. Further, to study annual variations in arrivals and prices of foodgrains and oilseeds the measures like standard deviation and coefficient of variation are used. The estimates of monthly average arrivals and prices, range, standard deviation and coefficient of variation have been calculated separately for each agricultural commodity.

The discussion made in this chapter is divided into two sections. Section 'A' deals with the annual variation in arrivals and prices of foodgrains and section 'B' deals with the annual variation in arrivals and prices of oilseeds.

SECTION A : ANNUAL VARIATION IN ARRIVALS AND PRICES OF FOODGRAINS

This section deals with the annual variations in arrivals and prices of foodgrains, namely, wheat, bajra, kharif jowar, rabi jowar, tur and greengram in the regulated markets in Marathwada.

Table 1

Annual Arrivals and Average. Prices of Wheat at Regulated Markets in Marathwada

(Arrivals in Quintals)
(Price in Rs. per Quintal)

YEARS	MARKETS													
	AURANGABAD		*LASUR*		*JALNA*		*PARTUR*		*PARBHANI*		*MANWAT*		*BEED*	
	Arrivals	*Prices*	*Arrivals*	*Prices*	*Arrivals*	*Prices*	*Arrivals*	*Prices*	*Arrivals*	*Prices*	*Arrivals*	*Prices*	*Arrivals*	*Prices*
1981	2211.08	216.75[a]	647.92	194.17[a]	2824.42	205.00	783.92	211.33	1063.08	202.75[a]	644.42	204.67	1237.42	209.08
1982	2486.25	224.50	618.75[a]	223.75	3071.25	222.17	446.17	212.17	1099.08	222.58	605.17	218.92	1155.00	228.83
1983	828.50	261.00	629.42	262.83	1841.58[a]	271.72	301.75	207.50	873.08	266.50	446.67	264.92	986.67	277.92
1984	4296.50	221.75	2205.83	196.92	7064.50[b]	204.92[a]	893.42[b]	197.25[a]	1259.08[b]	209.58	1081.50	203.67[a]	2490.08[b]	205.83[a]
1985	1676.67	241.08	815.75	247.08	3734.67	246.33	471.17	234.83	900.67	244.25	610.50	234.83	1892.00	242.08
1986	57[illegible].08	276.08	795.33	278.75	2435.58	295.50	178.67	294.25	406.75	290.42	213.08	278.50	956.92	291.67
1987	421.00[a]	276.58	756.25	271.17	3106.08	294.75	246.67	279.42	186.33	288.33	177.33	283.42	512.67[a]	300.00
1988	2152.08	301.60	1263.75	312.50	4998.25	317.92	409.92	316.83	184.17	336.67	121.50[a]	324.67	890.00	338.33
1989	4193.08	291.42	2350.75	301.17	6166.42	330.25	823.50	279.25	677.67	353.75	547.58	321.67	2171.75	317.17
1990	3736.00	305.33	1701.67	331.92	3517.50	325.42	497.92	298.58	499.25	316.33	1340.75[b]	296.08	1451.50	326.17
1991	3383.75	378.92	1490.58	432.58	3452.92	402.58	356.00	400.58	497.25	409.58	322.75	402.25	1091.83	438.42
1992	2868.83	496.92	1881.50	484.75	2581.58	530.00	230.50	560.08	420.00	503.58	281.08	524.73	1067.25	549.75
1993	3587.58	429.92	3772.83[b]	432.83	6146.42	452.42	453.50	467.25	248.75	484.42	409.00	454.58	1161.08	507.42
1994	5803.33[b]	492.92	2810.08	493.08	5064.33	514.50	442.17	538.50	114.50	535.25[b]	333.75	543.58	2188.83[b]	550.50
1995	2402.92	525.50[b]	1633.50	532.08[b]	2591.42	547.33[b]	161.08[a]	590.50[b]	44.33[a]	435.42	217.92	562.83[b]	842.92	581.25[b]
Average	2707.91	329.35	1558.26	333.04	3906.46	344.05	446.42	339.22	564.93	339.96	490.20	341.29	1339.69	357.63
Range between maximum and minimum	5382.33	308.75	3154.08	337.91	5222.92	342.41	732.34	393.25	1214.75	332.50	1219.25	359.16	1977.41	375.42
S.D.	1620.71	103.92	897.83	109.06	1537.06	113.86	220.50	132.67	378.84	106.39	328.36	121.09	557.95	127.84
C.V.	59.85	31.55	57.62	32.75	39.35	33.09	49.39	39.11	67.06	31.29	66.99	35.48	41.65	35.75

(contd.

Years	MARKETS													
	MANJLEGAON		LATUR		UDGIR		OSMANABAD		KALLAM		NANDED		NAIGAON	
	Arrivals	Prices	Arrivals	Prices	Arrivals	Prices	Arrivals	Prices	Arrivals	Prices	Arrivals	Prices	Arrivals	Prices
1981	883.17[b]	207.08[a]	1325.42	215.17	389.75	209.58	263.50	208.33[a]	236.75	223.75	890.17	212.00	390.33	180.75[a]
1982	414.83	222.83	3092.75	231.75	598.75	223.75	235.08	230.92	242.17	209.58	2107.17	216.83	666.25	200.17
1983	262.67	279.17	2043.75	280.00	343.00	270.25	173.25	283.83	324.92	260.83	1513.67	253.92	483.75	232.33
1984	554.17	208.25	6712.42[b]	207.08[a]	1530.92[b]	200.67[a]	518.75	216.00	190.67	206.42[a]	3325.25[b]	202.75[a]	1017.67[b]	186.67
1985	410.08	244.08	1867.25	245.67	566.25	238.17	131.33	267.50	379.83	269.50	1685.58	225.17	225.83	227.67
1986	303.58	286.92	1653.42	288.33	405.17	262.92	147.50	294.75	353.08	274.00	567.75	259.17	101.33	231.08
1987	265.33	297.75	1601.17	292.50	76.83	262.83	111.08	308.92	98.00[a]	305.00	316.17[a]	257.08	38.25[a]	196.75
1988	189.00[a]	328.67	1689.92	333.75	66.25	340.83	89.58[a]	339.08	342.33	304.58	580.08	307.08	123.33	277.33
1989	615.00	311.08	5390.08	306.67	410.42	271.42	288.00	323.50	801.17	288.50	2147.83	306.67	674.50	280.75
1990	496.00	306.58	3622.50	303.75	209.83	304.42	241.83	321.83	627.17	289.33	1914.25	294.50	319.67	244.58
1991	341.83	413.92	1833.33	377.92	50.33	396.25	187.92	430.08	397.50	401.33	1739.58	385.17	236.92	336.67
1992	541.42	488.58	518.92	540.42	14.58	493.50	208.25	547.75	189.00	554.92	2650.42	480.33	126.42	427.08
1993	640.58	473.00	507.33	491.25	23.50	435.92	161.75	522.33	396.25	473.08	2979.92	438.67	125.00	322.42
1994	534.08	545.17	1161.25	522.08	40.75	517.92[b]	721.17[b]	516.83	1111.42[b]	483.75	1804.92	493.50	255.58	449.17[b]
1995	505.00	555.50[b]	198.00[a]	630.42[b]	7.83[a]	479.17	257.08	586.08[b]	318.50	610.00[b]	1362.17	501.17[b]	81.33	402.58
Average	463.78	344.57	2214.50	351.12	315.61	327.17	249.07	359.85	400.58	343.64	1705.66	322.27	324.41	279.73
Range between maximum and minimum	694.17	348.42	6514.42	423.34	1523.09	317.25	631.59	377.75	1013.42	403.58	3009.08	298.42	979.42	268.42
S.D.	173.79	115.99	1752.83	127.80	382.12	105.68	160.22	123.04	254.03	124.55	851.01	104.87	268.58	85.95
C.V.	37.47	33.66	79.15	36.40	121.07	32.30	64.33	34.19	63.42	36.24	49.89	32.54	82.79	30.73

a = Minimum level;. b = Maximum level;. S.D. = Standard deviation;. C.V. = Coefficient of variation

Source: Data are compiled from the office records of selected markets.

Wheat

Wheat is the most important rabi crop grown in India as well as in Marathwada. The results presented in table 1 show that the range between the maximum and minimum annual arrivals of wheat is the highest in case of Latur (6514.42 quintals) market and the lowest in Osmanabad (631.59 quintals) market. As far as annual variation in prices is concerned, it is the highest in Latur (Rs. 423.34) market and the lowest in Naigaon (Rs. 268.42) market. Latur is a super market and the size of market arrivals is relatively higher in this market.

The standard deviation and coefficient of variation in arrivals of wheat are found to be the highest in Latur (1752.83) and Udgir (121.07) markets respectively and the lowest in Osmanabad (160.20) and Manjlegaon (37.47) markets respectively. It means the arrivals of wheat have varied to a greater extent in Latur and Udgir markets and to a smaller extent in Osmanabad and Manjlegaon markets during the study period. This may be due to the fact that Latur and Udgir markets are relatively larger sized markets than the Osmanabad and Manjlegaon markets. This can also be due to higher yield variability of wheat crop in the above market areas caused by natural factors. Wheat prices have exhibited larger variation in Partur market due to demand and supply conditions and low level of wheat arrivals. In case of annual price variation both the standard deviation and coefficient of variation are found to be the highest in Partur (132.67 and 39.11 respectively) market and the lowest in Naigaon (85.95 and 30.73 respectively) market. It means the price of wheat had exhibited relatively higher monthly variation in Partur market and relatively smaller variation in Naigaon market.

Bajra

Bajra is the most important kharif crop grown in Marathwada. The results presented in table 2 about this commodity reveal that the range between the maximum and the minimum annual arrivals is the highest in Jalna (18564.50 quintals) and the lowest in Kallam (403.66 quintals) market. As far as the range in annual prices is concerned, it is the highest in Kallam (Rs. 398.84) and the lowest in Manwat (Rs. 235.50) market.

Table 2
Annual Arrivals and Average Prices of Bajra at Regulated Markets in Marathwada

(Arrivals in Quintals)
(Price in Rs. per Quintal)

Years	MARKETS											
	AURANGABAD		LASUR		JALNA		PARTUR		MANWAT		BEED	
	Arrivals	Prices	Arrivals	Prices	Arrivals	Prices	Arrivals	Prices	Arrivals	Prices	Arrivals	Prices
1981	1083.75[a]	144.08	1291.67	136.67	1735.08[a]	150.50	372.58	118.08	296.58	97.00[a]	2370.83	171.08
1982	1970.00	130.25[a]	1314.67	137.58	4372.50	123.00	504.25	107.17	487.75[b]	121.92	3940.75	143.58
1983	3160.42	152.25	2075.58	157.42	4273.42	143.67	392.50	133.08	279.08	130.00	2548.25	163.75
1984	1777.92	134.17	1428.92	128.50[a]	3364.00	122.67[a]	241.58	104.42[a]	130.42	108.42	2024.75[a]	143.33[a]
1985	2581.42	234.75	2159.92	175.92	4383.50	155.83	178.75	139.33	88.25	141.25	3702.08	151.83
1986	2128.25	192.67	2500.67	205.83	7197.08	197.92	301.83	164.83	247.08	136.08	3095.50	200.75
1987	3993.08	179.42	5883.25[b]	179.00	20299.58[b]	180.00	636.17	166.50	463.17	164.17	6644.83	172.50
1988	3782.67	186.50	5234.08	174.83	7577.92	181.25	250.08	160.17	81.42	157.17	7547.75	187.50
1989	1809.50	159.42	1682.08	180.42	5470.17	171.17	139.75[a]	169.58	97.92	183.92	3107.58	214.58
1990	3181.50	196.92	1105.83	247.17	5824.50	197.08	187.17	149.83	58.83	132.50	5566.92	194.08
1991	1456.75	296.00	911.25	419.83	4926.17	299.42	142.92	269.83	26.92[a]	243.17	3779.17	317.42
1992	3017.92	394.42	2601.00	427.67	13830.83	395.83	814.17[b]	341.25	466.25	332.50[b]	4894.17	413.75
1993	5362.92[b]	158.00	3564.17	227.25	13909.25	219.25	531.50	212.92	340.83	238.50	10962.50[b]	230.25
1994	3344.83	357.50	1622.17	368.08	6277.50	371.58	405.42	309.42	217.08	290.67	6345.42	384.42
1995	2030.08	471.75[b]	827.83[a]	488.25[b]	7038.25	454.25[b]	667.58	438.92[b]	282.67	292.25	4682.42	498.17[b]
Average	2712.07	225.87	2280.21	243.63	7365.32	224.23	384.42	199.02	237.62	184.63	4747.53	239.13
Range between maximum and minimum	4279.71	341.50	5055.42	359.75	18564.50	331.58	674.42	334.50	460.83	235.50	8937.75	354.84
S.D.	1247.83	101.82	1468.59	143.32	4763.18	101.64	201.58	94.53	150.15	72.83	2299.51	107.20
C.V.	46.01	45.08	64.41	58.83	64.67	45.33	52.44	47.50	63.19	39.44	48.44	44.83

contd.

Years	MARKETS							
	MANJLEGAON		LATURQ		UDGIR		KALLAM	
	Arrivals	Prices	Arrivals	Prices	Arrivals	Prices	Arrivals	Prices
1981	2702.00	152.67	4723.92	153.75	569.08	149.42	8.58	131.00
1982	3387.92[b]	138.33	4500.75	119.08[a]	643.42	121.92[a]	7.08	124.00
1983	1203.33	152.42	2433.42	149.00	194.08	145.25	3.17[a]	103.08
1984	646.67	136.08[a]	1162.25[a]	120.50	167.92	123.83	4.92	51.08[a]
1985	946.17	146.17	3795.67	154.83	377.92	145.92	9.00	127.83
1986	1135.83	190.83	2869.67	192.08	299.17	179.67	18.08	120.92
1987	2309.17	171.17	9075.25[b]	170.83	739.08[b]	166.75	275.33	169.67
1988	1112.17	179.25	7089.08	162.50	363.75	160.17	406.83[b]	148.17
1989	1043.75	200.83	2385.75	173.75	254.92	168.92	217.42	161.83
1990	834.75	190.17	2001.75	176.67	83.42	171.25	39.33	161.92
1991	596.00[a]	309.25	1857.58	278.75	51.58[a]	259.08	27.50	264.33
1992	2372.75	372.42	2071.58	377.08	171.50	290.42	219.58	367.25
1993	3255.67	235.83	3546.42	212.50	150.92	204.50	367.17	205.25
1994	1200.33	381.00	6415.08	348.33	122.58	282.67	358.67	340.25
1995	2374.67	485.00[b]	6808.75	454.17[b]	140.83	407.75[b]	293.67	449.92[b]
Average	1674.75	229.43	4049.13	216.25	288.68	198.50	150.42	195.10
Range between maximum and minimum	2791.92	348.92	7913.00	335.09	687.56	285.83	403.66	398.84
S.D.	922.99	103.50	2261.28	97.83	204.32	76.12	152.75	107.70
C.V.	55.11	45.11	55.85	45.24	70.78	38.35	101.55	55.20

a = Minimum level;. b = Maximum level;. S.D. = Standard deviation;. C.V. = Coefficient of variation

Source: As indicated in table 1.

The coefficient of variation in arrivals of bajra is found to be the highest in Kallam (101.56) market and the lowest in Aurangabad (46.01) market. Both standard deviation and coefficient of variation of annual prices of bajra are found to be the highest in Lasur (143.32 and 58.83 respectively) market and the lowest in Udgir (76.12 and 38.35 respectively) market. It means the arrivals of bajra have varied to a greater extent in Kallam market and to a smaller extent in Aurangabad market during the study period. The prices of bajra have shown larger annual variation in Lasur market and lesser annual variation in Udgir market. The larger variation in arrivals of bajra in case of Kallam market may be due to the fact that bajra is not a popular crop in surrounding area of this market. So the level of arrivals of this commodity has been very small in this market which ultimately resulted into greater variation in its arrivals. Lasur market showed higher annual price variation in case of bajra. It means the prices of bajra at Lasur market have remained more unstable due to year to year variations in arrivals caused by the natural factors.

Kharif Jowar

Kharif Jowar is the main kharif cereal crop of the Marathwada region. The results presented in table 3 indicate that the range between maximum and minimum annual arrivals of this commodity is the highest in case of Latur (23295.67 quintals) market and the lowest in case of Partur (552.33 quintals) market. The range in annual prices of kharif jowar is found to be the highest in case of Beed (Rs. 330.80) market and the lowest for Lasur (Rs. 241.75) market. The coefficient of variation of arrivals of kharif jowar is found to be the highest in case of Lasur (75.09) market and the lowest in case of Latur (33.93) market. It means that the annual variation of arrivals of kharif jowar has been relatively higher in Lasur market and relatively lower in Latur market during the study period.

As far as the annual variation in its prices is concerned, it is found that the values of standard deviation and coefficient of variation are the highest for Beed (97.53) and Aurangabad (50.19) markets and the lowest for Partur (73.86 and 42.32 respectively) market. It means that the annual variation in prices of kharif jowar has been relatively higher in Beed and Aurangabad markets and relatively smaller in Partur market during the study period.

Table 3

Annual Arrivals and Average. Prices of Kharif Jowar at Regulated Markets in Marathwada

(Arrivals in Quintals)
(Price in Rs. per Quintal)

Years	MARKETS											
	AURANGABAD		LASUR		JALNA		PARTUR		PARBHANI		BEED	
	Arrivals	Prices	Arrivals	Prices	Arrivals	Prices	Arrivals	Prices	Arrivals	Prices	Arrivals	Prices
1981	1932.58	153.75	1294.50	104.92[a]	2479.67	146.00	608.50	120 92	928.67	137.42	708.67	152.25
1982	2802.25	120.58	1176.00	108.83	6653.83	119.75[a]	726.00	106.33[a]	2061.58[b]	106.75	726.83	124.75[a]
1983	4337.33[b]	144.17	1145.50	126.67	9611.83	142.58	557.00	123.00	1343.50	114.83	792.00	142.17
1984	2309.83	140.08	1163.83	134.33	4576.33	143.08	569.17	129.25	1161.08	108.50	1099.67	146.25
1985	3054.42	134.58	1974.67	143.33	7079.25	132.25	587.67	116.92	1253.00	103.83	1380.42[b]	134.33
1986	2929.33	153.50	1333.83	157.67	3685.00	161.75	328.75	152.00	580.75	128.25	1010.25	147.08
1987	3809.08	156.33	2431.50[b]	149.42	10262.00[b]	155.25	725.08	142.92	515.58	103.75[a]	1324.67	153.75
1988	1818.75	121.25	559.58	120.50	3130.83	127.50	264.58	124.92	378.00	130.42	383.50	136.25
1989	1934.83	182.50	610.67	181.58	2781.42	196.75	377.50	142.25	67.50[a]	172.08	444.25	237.08
1990	3293.00	145.92	440.58	206.42	3604.58	151.50	248.08	158.92	203.92	179.17	249.17	170.50
1991	1449.17	216.17	348.92	346.67[b]	3371.25	244.83	291.00	216.33	247.58	240.00	246.83	229.33
1992	830.08[a]	383.67	494.17	298.00	5394.58	387.08	800.33[b]	380.25[b]	1034.50	368.00	360.83	402.58
1993	1437.00	101.75[a]	192.33	152.92	5366.33	160.58	579.83	174.50	1045.83	162.75	122.75	184.67
1994	2473.75	329.92	85.92[a]	199.42	1709.58[a]	277.17	248.00[a]	246.33	708.25	188.75	87.25[a]	277.92
1995	1054.42	421.08[b]	111.75	345.00	3975.17	438.42[b]	583.92	283.17	200.83	370.00[b]	91.58	455.58[b]
Average	2364.39	193.68	890.92	185.05	5245.44	198.97	499.69	174.53	782.04	174.30	601.91	206.30
Range between maximum and minimum	3507.25	319.33	2345.58	241.75	8552.42	318.67	552.33	273.92	1994.08	266.25	1293.17	330.83
S.D.	1104.57	97.21	668.98	78.45	2587.23	94.13	183.17	73.86	530.48	84.96	427.57	97.53
C.V.	46.72	50.19	75.09	42.39	49.32	47.31	36.66	42.32	67.83	48.75	71.03	47.27

contd.

Years	MARKETS									
	LATUR		UDGIR		OSMANABAD		KALLAM		NAIGAON	
	Arrivals	*Prices*	*Arrivals*	*Prices*	*Arrivals*	*Prices*	*Arrivals*	*Prices*	*Arrivals*	*Prices*
1981	15435.42	138.83	6150.92	138.67	911.58	143.75	2145.92	155.92	829.00	138.83
1982	11463.17	112.42[a]	6289.33	102.58[a]	1063.17	115.00[a]	2488.58	113.67[a]	1024.50	93.42
1983	18109.92	121.42	9329.17[b]	115.58	956.42	131.42	1478.75	117.17	909.42	101.17
1984	8152.75[a]	144.17	1086.58[a]	126.92	869.42	139.25	1422.83	138.25	392.92[a]	124.00
1985	18388.92	112.50	4188.08	108.33	1788.00[b]	130.83	3885.67[b]	138.17	1324.83	99.92
1986	13367.08	142.92	3553.42	130.08	1132.92	152.83	1472.50	160.75	1702.17[b]	127.58
1987	19215.42	165.92	2393.25	145.17	424.42	162.33	1256.83	172.50	1040.83	141.67
1988	31448.42[b]	150.17	7128.25	126.83	213.58[a]	150.33	2280.83	129.83	715.75	108.75
1989	11503.42	227.25	2156.50	219.58	594.33	208.42	1400.00	245.42	454.33	232.17
1990	12004.17	165.00	3039.42	144.83	504.25	159.58	1863.42	151.42	546.17	126.67
1991	10313.58	234.58	2978.58	202.58	729.50	217.17	1633.75	225.58	643.08	188.75
1992	14119.25	388.75	5921.50	346.50	938.83	391.92	2529.25	394.92	1192.25	351.67[b]
1993	17106.00	177.92	8437.92	156.58	767.83	177.50	2283.17	167.33	429.33	119.17
1994	17072.33	265.83	5869.75	258.58	824.08	261.33	1589.67	267.92	1134.50	247.08
1995	16537.75	442.92[b]	4080.58	365.92[b]	520.08	416.50[b]	814.58[a]	442.50[b]	1376.58	282.00
Average	15615.84	199.37	4840.22	179.25	815.89	197.21	1903.05	201.42	914.38	165.52
Range between maximum and minimum	23295.67	330.50	8242.59	263.34	1574.42	301.50	3071.09	328.83	1309.25	258.25
S.D.	5298.25	95.87	2324.12	81.10	357.13	89.27	713.47	96.14	376.94	75.20
C.V.	33.93	48.09	48.02	45.24	43.77	45.27	37.49	47.73	41.21	45.43

a = Minimum level;. b = Maximum level;. S.D. = Standard deviation;. C.V. = Coefficient of variation

Source: As indicated in table 1.

From the above discussion it is obvious that the arrivals of kharif jowar have exhibited greater annual variation in Lasur market but the prices of this commodity have exhibited greater variation in Beed and Aurangabad markets because the production of kharif jowar depends upon the favourableness of the natural factors such as rainfall and climate. These factors are unfavourable in Lasur market area. More annual variation in kharif jowar prices in Aurangabad and Beed markets is due to the size of markets.

Rabi Jowar

Rabi jowar is an important crop of the Marathwada region. In the Maharashtra state, rabi jowar is grown mainly in the Marathwada region and some parts of western Maharashtra. The results presented in table 4 show that the range between maximum and minimum annual arrivals is the highest in case of Jalna (10908.00 quintals) market and the lowest in case of Kallam (544.16 quintals) market. In case of rabi jowar, the price range is found to be the highest for Lasur (Rs. 525.17) market and lowest for Nanded (Rs. 290.08) market.

The annual variations in arrivals and prices of rabi jowar are also measured by way of standard deviation and coefficient of variation in arrivals and prices. The standard deviation and coefficient of variation in arrival are found to be highest in case of Jalna (2860.54) and Udgir (152.53) markets respectively and the lowest in case of Kallam (146.67 and 31.83 respectively) market. The higher degree of variation in arrivals at Jalna and Udgir markets may be attributed to larger variations in acreage and yield of this crop during the study period. But the interesting thing to note here is that both standard deviation and coefficient of variation in annual prices of rabi jowar are found to be the highest in case of Lasur (153.35 and 48.47 respectively) market and the lowest for Nanded (88.09) and Parbhani (37.53) markets respectively. It means that the annual prices of rabi jowar have been subject to lesser variation in Nanded and Parbhani markets during the study period. On the contrary, they have varied to a larger extent in Lasur market during the period of this study. These larger annual variations in prices of rabi jowar at Lasur market may be due to greater fluctuations in arrivals caused by natural factors.

Table 4

Annual Arrivals and Average Prices of Rabi Jowar at Regulated Markets in Marathwada

(Arrivals in Quintals)
(Price in Rs. per Quintal)

Years	MARKETS													
	AURANGABAD		LASUR		JALNA		PARTUR		PARBHANI		MANWAT		BEED	
	Arrivals	Prices	Arrivals	Prices	Arrivals	Prices	Arrivals	Prices	Arrivals	Prices	Arrivals	Prices	Arrivals	Prices
1981	2827.17	201.58	3748.92	153.50[a]	4902.83	187.67	1240.83	179.50	1434.42	196.25	1662.67	179.08	2739.08	188.33
1982	3123.17	178.33[a]	4541.67	154.17	5987.17	169.08[a]	2427.92[b]	159.00[a]	1913.50[b]	177.50[a]	2386.00[b]	166.00[a]	3336.58	171.92[a]
1983	1941.08	204.25	2653.50	194.00	5534.08	204.08	1895.92	196.67	1285.50	210.25	1013.83	197.17	2613.75	202.83
1984	2305.50	213.25	3796.75	197.00	5717.83	212.50	944.25	195.83	420.92	225.00	787.50	202.08	2557.33	209.08
1985	1750.83	208.75	1474.75[a]	217.67	7034.92	196.25	1337.58	178.08	1063.92	190.33	1550.92	171.33	5067.00	189.17
1986	1068.42	237.83	1679.33	229.42	4528.83[a]	248.42	578.33[a]	230.33	798.92	232.25	1444.00	224.75	3681.42	228.75
1987	1005.33[a]	238.42	2861.42	215.67	10264.25	225.00	1495.25	184.58	700.67	220.83	1952.08	204.92	2806.25	210.42
1988	3339.00[b]	271.83	6092.17[b]	248.67	15436.83[b]	257.92	1653.58	227.33	778.25	253.75	1808.08	246.58	7249.83[b]	253.33
1289	1990.08	326.42	4563.42	311.50	8574.67	337.25	843.25	317.83	247.00	366.67	591.83	347.92	4006.42	333.75
1990	2611.67	259.50	4520.25	311.25	9915.92	254.17	855.50	251.83	223.17	292.92	778.67	256.75	4645.33	245.50
1991	1425.42	401.92	2493.25	505.75	6616.83	405.00	845.25	394.67	299.42	424.58	656.75	401.75	4092.67	401.17
1992	1354.42	551.58	1877.17	562.33	6468.83	536.42	608.00	528.08	495.42	551.33[b]	458.58	517.00	2461.50[a]	557.33
1993	2017.25	360.58	3189.33	351.33	11938.50	346.67	768.17	328.50	456.50	389.33	1239.33	326.67	4945.58	346.25
1994	3215.08	405.08	3242.08	415.17	7702.83	420.17	591.67	406.58	266.17	424.17	860.08	406.92	4050.33	432.58
1995	1784.50	635.67[b]	2639.67	678.67[b]	8748.92	629.58[b]	698.58	606.50[b]	79.08[a]	478.17	402.90[a]	621.92[b]	3435.33	652.25[b]
Average	2117.26	313.00	3291.58	316.41	7958.22	308.68	1118.94	292.35	697.52	308.89	1172.82	298.06	3845.89	308.18
Range between maximum and minimum	2333.67	457.34	4617.42	525.17	10908.00	460.50	1849.59	447.50	1834.42	373.83	1984.00	455.92	4788.33	480.33
S.D.	880.18	131.36	1226.53	153.35	2860.54	132.15	528.47	132.44	506.55	115.94	581.87	132.7	1235.87	140.83
C.V.	41.57	41.97	37.26	48.47	35.94	42.81	47.23	45.30	72.62	37.53	49.61	44.53	32.13	45.70

contd.

Years	MARKETS MANJLEGAON		LATUR		UDGIR		OSMANABAD		KALLAM		NANDED	
	Arrivals	Prices	Arrivals	Prices	Arrivals	Prices	Arrivals	Prices	Arrivals	Prices	Arrivals	Prices
1981	4106.08	178.00	1923.67	203.83	814.25[b]	204.83	415.83	207.08	556.17	210.33	634.42[a]	172.58
1982	7050.67[b]	171.42[a]	667.75	193.08[a]	147.00	200.33	363.67	194.58	717.42	185.75[a]	1369.42	137.42
1983	3617.58	198.50	1072.67	213.17	233.50	222.25	409.50	208.92	440.58	209.33	1814.17	140.58
1984	1581.00	202.00	2086.25	224.17	42.08	248.17	395.67	218.50	437.50	207.33	1525.75	154.08
1985	3862.08	184.83	2608.67	194.08	253.33	199.67	723.33	183.67[a]	750.08[b]	201.08	3996.50	118.58[a]
1986	2172.33	225.08	1189.33	245.83	158.75	238.42	478.33	234.08	502.83	233.25	2226.17	141.25
1987	3411.33	204.92	3232.33	233.75	101.58	240.92	392.92	232.00	205.92[a]	236.00	1857.17	168.08
1988	4068.67	251.42	3695.00[b]	283.33	21.33	218.92	168.17[a]	274.92	396.17	257.50	928.83	162.58
1289	1999.42	335.58	3177.67	359.58	8.33[a]	185.08[a]	254.33	362.08	349.17	342.08	1202.08	272.08
1990	1822.08	244.50	1701.83	282.08	24.08	281.58	326.00	284.83	361.67	262.33	1923.33	170.92
1991	1118.33[a]	404.42	1186.83	423.33	11.00	340.92	331.58	422.00	410.33	413.83	3120.087	274.83
1992	2030.58	519.25	502.50	568.75	11.42	560.17	622.50	545.17	267.83	553.08	1635.92	389.92
1993	3596.33	336.67	502.58	387.92	81.08	355.25	581.50	332.67	383.33	328.25	1498.42	196.42
1994	1489.92	450.75	698.67	463.75	26.33	447.92	662.00	421.25	538.17	445.83	5944.67[b]	265.92
1995	1453.50	626.33[b]	456.00[a]	685.83[b]	23.42	646.75[b]	948.33[b]	629.00[b]	595.42	626.67[b]	3477.83	408.66[b]
Average	2891.99	302.24	1646.78	330.83	130.50	306.08	471.58	316.72	460.84	314.18	2210.32	211.59
Range between maximum and minimum	5932.34	454.91	3239.00	492.75	805.92	461.67	780.16	445.33	544.16	440.92	5310.25	290.08
S.D.	1516.30	135.50	1061.30	144.51	199.05	136.42	195.35	130.82	146.67	132.67	1343.85	88.09
C.V.	52.43	44.83	64.45	43.68	152.53	44.57	41.43	41.31	31.83	42.23	60.80	41.63

a = Minimum level;. b = Maximum level;. S.D. = Standard deviation;. C.V. = Coefficient of variation

Source: As indicated in table 1.

Tur

Tur is the principal pulse crop of Marathwada. In case of tur, the table 5 reveals that the range between the maximum and minimum annual arrivals is the highest in case of Latur (16971.00 quintals) market and the lowest for Naigaon (323.41 quintals) market. But the price range is found to be the highest in case of Udgir (Rs. 1346.42) market and the lowest for Naigaon (Rs. 509.42) market. The standard deviation and coefficient of variation in annual arrivals are found to be the highest in case of Latur (4304.86) and Aurangabad (69.41) markets respectively and the lowest for Naigaon (87.34) and Latur (33.77) markets respectively. As far as annual price variation is concerned the standard deviation and coefficient of variation are the highest for Latur (365.80) and Lasur (53.66) markets respectively and the lowest for Naigaon (162.83 and 32.26 respectively) market. Both the standard deviation and coefficient of variation have indicated that the annual prices of tur commodity have been subject to lesser variation in Naigaon market. They have varied to a greater extent in Latur and Lasur markets. Annual arrivals of tur have shown greater stability in Naigaon and Latur markets and lesser stability in Latur and Aurangabad markets during the study period. In case of tur higher variation of arrivals is found in Latur and Aurangabad markets. This may be due to larger size of these markets and natural factors.

Greengram

Greengram is the important short duration pulse crop grown in Marathwada. In case of greengram the range between maximum and minimum annual arrivals is found to be the highest in case of Jalna (12500.66 quintals), followed by Parbhani (6534.00 quintals) and Latur (5072 quintals) markets and the lowest in case of Kallam (183.75 quintals) market. But the range in annual prices is the highest for Latur (Rs. 1023.33) market and the lowest for Partur (616.17) market.

The standard deviation of annual arrivals of greengram is found to be the highest for Jalna (3497.98) market and the lowest for Kallam (52.60) market. But the value of coefficient of variation in annual arrivals of this commodity is found to be the highest for Aurangabad (71.51) and the lowest for Partur (35.98) markets. From

Tables 5
Annual Arrivals and Average Prices of Tur at Regulated Markets in Marathwada

(Arrivals in Quintals)
(Price in Rs. per Quintal)

Years	MARKETS													
	AURANGABAD		LASUR		JALNA		PARTUR		PARBHANI		MANWAT		BEED	
	Arrivals	Prices	Arrivals	Prices	Arrivals	Prices	Arrivals	Prices	Arrivals	Prices	Arrivals	Prices	Arrivals	Prices
1981	384.83[a]	294.25[a]	309.25	260.42[a]	2057.83[a]	338.33[a]	1314.42	363.25	1856.67	316.00[a]	1348.67	339.83	1097.00	336.50[a]
1982	886.83	344.42	279.33	322.08	3852.42	355.75	1153.75	337.67[a]	1761.75	328.92	2035.50	251.75[a]	1354.08	337.92
1983	782.17	430.42	174.92[a]	311.25	3205.42	451.08	610.00[a]	423.08	1316.75	383.42	743.75[a]	422.58	978.50	442.00
1984	518.58	393.83	541.75	411.50	7654.67	427.17	1733.17	462.00	3211.17	429.08	3338.00	422.50	2266.25	428.17
1985	2015 25	366.58	534.75	349.25	8779.83	376.08	1836.92	354.33	3717.67[b]	359.75	3561.50	354.67	3369.33	360.92
1986	1704.00	440.58	653.50	340.33	7696.42	458.00	1341.33	446.42	2890.67	440.58	2893.08	436.92	2010.50	454.58
1987	1317.75	508.25	622.33	577.33	2879.75	638.50	660.33	540.92	1744.42	612.08	2222.75	555.42	936.17[a]	634.58
1988	2084.83	653.75	952.17	600.25	11259.73[b]	676.58	1786.00	653.33	3291.17	660.42	3935.00[b]	648.42	2846.58	674.17
1289	165[illegible].42	620.50	2450.58[b]	595.58	10544.50	624.42	1728.92	615.42	2555.75	627.75	1892.08	623.67	4840.67[b]	623.92
1990	3803.00	770.58	2154.08	820.83	8852.92	793.08	1580.58	775.50	2695.17	771.67	2138.58	700.83	3629.42	769.25
1991	4540.17	984.83	1412.00	1043.75	7411.17	1011.25	1374.08	985.42	1754.67	960.42	1623.33	984.50	3156.50	1037.83
1992	2113.67	838.83	1198.00	1009.33	3765.17	1048.42	767.33	932.42	1312.17	1001.33	884.33	861.67	1435.67	1051.83
1993	3865.33	964.83	1298.25	881.50	7631.50	963.58	1269.08	929.25	2769.67	892.67	2541.50	911.42	1857.67	957.67
1994	5721.58[b]	1049.33	1287.83	1006.42	8209.58	1081.67	3924.50[b]	969.58	2606.67	1077.25	3235.67	1056.33	3066.50	1060.08
1995	4266.83	1564.67[b]	739.25	1568.67[b]	5550.83	1614.67[b]	1004.50	1394.42[b]	1028.08[a]	1475.25[b]	1607.58	1525.25[b]	1318.25	1659.08[b]
Average	2377.42	681.71	973.87	673.23	6623.45	723.89	1472.33	678.87	2300.83	689.11	2266.75	673.05	2277.54	721.90
Range between maximum and minimum	5336.75	1270.42	2275.66	1308.25	9201.92	1276.34	3314.50	1056.75	2689.59	1159.25	3191.25	1273.50	3904.50	1322.58
S.D.	1550.23	338.76	645 15	361.28	2785.67	349.24	761.73	298.71	789.95	326.47	936.23	331.80	1123.03	359.73
C.V.	69.41	49.69	66.25	53.66	42.06	48.24	51.74	44.00	34.33	47.38	41.30	49.30	49.31	49.83

contd.

Years	MARKETS													
	MANJLEGAON		*LATUR*		*UDGIR*		*OSMANABAD*		*KALLAM*		*NANDED*		*NAIGAON*	
	Arrivals	*Prices*	*Arrivals*	*Prices*	*Arrivals*	*Prices*	*Arrivals*	*Prices*	*Arrivals*	*Prices*	*Arrivals*	*Prices*	*Arrivals*	*Prices*
1981	751.17[a]	339.00	8968.92	347.92[a]	2544.17	348.83[a]	511.58	351.67[a]	543.75	338.25[a]	1287.17	348.08[a]	152.67	291.75[a]
1982	2359.08	331.58[a]	14512.17	381.08	4315.42	368.67	937.17	379.92	1176.25[b]	366.50	1609.00	366.42	169.17	306.17
1983	1305.08	404.83	10765.42	467.92	5527.58	458.58	635.17	461.33	530.33	441.58	1359.83	450.17	120.17	296.92
1984	3121.17	435.17	11915.16	446.25	4151.92	441.58	515.83	443.75	305.00	371.83	1587.92	422.08	96.50	380.75
1985	2910.75	354.42	16462.33	374.17	7960.75	376.17	634.00	374.17	1018.42	360.58	1894.25	364.42	207.75	348.42
1986	2928.58	447.58	15173.83	460.00	8381.67	451.17	604.50	452.58	948.17	451.17	3074.25	451.08	168.50	355.08
1987	2234.67	580.67	5050.00[a]	676.25	1467.58[a]	675.92	447.58[a]	594.42	220.92[a]	584.33	1252.25	656.25	104.33	427.92
1988	3152.67	654.17	22021.00[b]	707.08	10218.17[b]	703.17	954.75	673.83	1165.58	638.33	3745.67[b]	673.33	409.33[b]	589.50
1989	3739.33[b]	618.83	15293.17	650.83	4636.17	644.42	467.42	629.50	989.08	611.25	1510.33	617.25	199.50	605.67
1990	1994.42	766.00	17129.17	798.33	6060.17	813.08	489.67	812.08	792.67	755.08	1413.08	800.92	135.00	571.33
1991	1660.58	995.75	15934.00	1100.83	4109.83	1061.33	834.25	1047.83	782.67	931.75	1144.50	1005.33	85.92[a]	700.00
1992	1227.75	993.83	7239.50	1063.75	1597.25	1048.83	1038.58	1018.50	244.33	988.50	921.08	1017.00	152.67	645.83
1993	2135.83	949.58	8398.50	1001.25	5396.00	962.67	783.92	915.25	320.50	858.42	1774.58	915.00	215.42	576.67
1994	2929.33	1055.08	10464.50	1107.25	8070.00	1118.57	1472.83[b]	1065.25	784.75	1033.87	1574.00	1051.42	333.75	801.17[b]
1995	1108.58	1466.25[b]	11892.83	1688.33[b]	5560.83	1695.25[b]	890.00	1473.00[b]	230.17	1406.50[b]	851.33[a]	1522.92[b]	95.83	674.17
Average	2237.27	692.85	12748.03	751.42	5333.17	744.55	747.82	712.87	670.17	675.86	1666.62	710.78	176.43	504.76
Range between maximum and minimum	2988.16	1134.67	16971.00	1340.41	8750.59	1346.42	1025.25	1121.33	955.33	1068.25	2894.34	1174.84	323.41	509.42
S.D.	859.19	324.79	4304.86	365.80	2439.47	365.02	271.92	320.45	337.69	303.94	747.31	327.40	87.34	162.83
C.V.	38.40	46.88	33.77	48.68	45.74	49.03	36.36	44.95	50.39	44.97	44.84	46.06	49.50	32.26

a = Minimum level;. b = Maximum level;. S.D. = Standard deviation:. C.V. = Coefficient of variation

Source : As indicated in table 1.

Table 6

Annual Arrivals and Average Prices of Greengram At Regulated Markets in Marathwada

(Arrivals in Quintals)
(Price in Rs. per Quintal)

Years	MARKETS									
	AURANGABAD		JALNA		PARTUR		PARBHANI		MANWAT	
	Arrivals	Prices	Arrivals	Prices	Arrivals	Prices	Arrivals	Prices	Arrivals	Prices
1981	400.83	254.17	6006.92	371.17	425.25[a]	286.42	1323.83	264.67[a]	251.92[a]	261.75[a]
1982	155.08	250.42[a]	11849.00	305.50[a]	965.42	273.75[a]	3402.08	269.67	1525.33	268.75
1983	739.00	287.08	16281.83[b]	346.00	697.50	313.42	5109.25	309.25	2056.25	318.00
1984	165.25	387.42	6848.17	424.08	684.00	388.50	7360.92[b]	407.58	542.25	388.75
1985	1318.75	414.92	13502.08	488.08	753.17	405.58	3534.92	460.58	1549.67	411.42
1986	929.75	318.83	7688.75	394.58	859.50	350.92	3929.50	399.08	1150.42	350.42
1987	1540.58	452.58	8992.67	458.58	1353.83	460.00	7252.92	456.92	3956.67[b]	447.58
1988	1651.75[b]	556.92	14864.58	502.50	1266.67	502.83	2335.08	571.33	1229.58	485.00
1989	512.17	579.00	7547.83	604.42	808.00	466.92	1376.33	492.08	810.33	482.58
1990	993.67	613.42	6048.25	607.58	912.92	531.58	826.92[a]	540.00	450.25	583.75
1991	565.67	589.00	7556.00	654.83	796.92	479.67	1248.42	644.17	610.08	546.67
1992	1082.25	905.08	11408.25	897.67	1527.83	621.67	3605.75	878.67	2114.25	775.50
1993	605.83	781.58	10973.17	915.33	1634.92[b]	889.92[b]	4791.17	924.17	2449.25	943.25
1994	36.08[a]	529.58	3781.17[a]	1054.58	528.83	836.50	884.42	858.83	564.08	902.58
1995	224.08	1165.33[b]	6913.92	1257.08[b]	1043.25	851.25	1888.00	1229.58[b]	1940.33	972.75[b]
Average	728.05	539.02	9350.87	618.80	950.53	510.60	3257.97	580.44	1413.38	542.58
Range between maximum and minimum	1615.67	914.91	12500.66	951.58	1209.67	616.17	6534.00	964.9	3704.75	711.00
S.D.	520.60	247.30	3497.18	276.04	342.00	196.82	2078.99	268.15	953.01	235.11
C.V.	71.51	45.88	37.40	44.61	35.98	38.55	63.81	46.27	67.43	43.33

contd.

Years	*MARKETS*							
	LATUR		*UDGIR*		*KALLAM*		*NANDED*	
	Arrivals	*Prices*	*Arrivals*	*Prices*	*Arrivals*	*Prices*	*Arrivals*	*Prices*
1981	1713.08	326.92	318.92	333.33	130.17	275.17	306.67	348.17
1982	4517.83	283.75[a]	1077.33	271.00	81.83	258.75[a]	953.83	304.75[a]
1983	4642.92	326.67	694.17	259.75[a]	205.58[b]	275.17	1514.17[b]	329.17
1984	2148.67	430.00	196.92[a]	330.92	58.00	330.25	804.75	433.50
1985	2987.83	451.67	812.75	429.33	21.85[a]	378.25	493.75	470.75
1986	1356.67	379.58	659.25	420.17	39.25	385.25	290.67	398.50
1987	5919.58[b]	452.92	1712.58	464.58	152.17	377.58	755.17	458.75
1988	5461.25	604.58	1924.50[b]	621.00	176.25	541.00	360.33	628.50
1989	2916.83	611.67	615.67	474.75	79.75	627.83	463.50	534.08
1990	2138.17	614.58	923.75	576.83	113.50	572.33	343.92	584.08
1991	2153.75	676.67	534.58	631.00	39.42	663.50	457.83	643.67
1992	2122.50	849.17	973.17	824.67	69.33	527.67	428.58	830.58
1993	2053.75	911.42	1396.50	960.92	115.08	808.67	691.00	970.00
1994	847.33[a]	1046.58	356.00	1089.75	80.58	806.67	221.00[a]	1036.25
1995	2416.67	1307.08[b]	1531.33	1270.92[b]	37.83	875.67[b]	390.33	1082.92[b]
Average	2893.12	618.22	915.16	597.26	93.37	513.58	565.03	603.58
Range between maximum and minimum	5072.25	1023.33	1727.58	1011.17	183.75	616.92	1293.17	778.17
S.D.	1474.23	286.37	505.91	298.39	52.60	201.58	325.44	251.61
C.V.	50.96	46.32	55.28	49.96	56.33	39.25	57.60	41.69

a = Minimum level;. b = Maximum level;. S.D. = Standard deviation;. C.V. = Coefficient of variation

Source: As indicated in table 1.

this it can be said that the arrivals of greengram have been subject to greater variation in Jalna and Aurangabad markets because Jalna and Aurangabad markets are super grade markets and in these market areas natural factors are favourable to grow greengram. On the contrary, the variation of annual arrivals of greengram has been of lesser degree in Kallam and Partur markets due to low production of greengram.

Both the standard deviation and coefficient of variation in annual prices of greengram are found to be the highest in case of Udgir (i.e. 298.39 and 49.96 respectively) market and the lowest in case of Partur (196.82 and 38.53 respectively) market. It means that annual prices of greengram have been most unstable in Udgir market and least unstable in Partur market during the study period. The higher degree of annual variations in arrivals of this commodity at Udgir market is due to natural factors.

SECTION B : ANNUAL VARIATION IN ARRIVALS AND PRICES OF OILSEEDS

This section deals with the annual variation in arrivals and prices of oilseeds at regulated markets in Marathwada. The commodity-wise analysis of annual variations in prices and arrivals is presented below :

Safflower

Safflower is an important oilseed crop grown in rabi season on medium to black soils in the Marathwada region. The results presented in the table 7 reveal that the range between the maximum and minimum annual arrivals of safflower is the highest in case of Jalna (4562.50 quintals) market and the lowest in case of Osmanabad (128.25 quintals) market. As far as annual prices are concerned, the range between the maximum and minimum is found to be the highest in case of Lasur (Rs. 756.59) market and the lowest for Osmanabad (Rs. 548.33) market. The coefficient of variation in annual arrivals of safflower is found to be the highest in case of Osmanabad (80.36) market and the lowest for Latur (31.62) market. It means the variation in annual arrivals has been of smaller degree in Latur market and it has been of a greater degree in Osmanabad market during the period of this study.

Table 7
Annual Arrivals and Average Prices of Safflower at Regulated Markets in Marathwada

(Arrivals in Quintals)
(Price in Rs. per Quintal)

Years	MARKETS											
	AURANGABAD		LASUR		JALNA		PARTUR		PARBHANI		MANWAT	
	Arrivals	Prices	Arrivals	Prices	Arrivals	Prices	Arrivals	Prices	Arrivals	Prices	Arrivals	Prices
1981	917.58	398.75	505.67	358.83[a]	3345.17	396.00	466.75	399.00	1011.42	363.92[a]	2637.25[b]	386.58[a]
1982	887.17	379.75[a]	532.00	371.08	2212.42	384.58[a]	570.83	357.17[a]	897.17	374.67	2019.25	388.42
1983	784.42	437.33	491.50	430.33	3884.42	436.92	448.67	385.33	573.42	417.33	2123.58	413.92
1984	1328.17	421.67	1311.08	443.67	4189.83	446.50	611.67	399.42	1187.00	437.67	2376.83	428.50
1985	1605.42	383.00	1326.33	385.00	5288.00[b]	384.92	695.42	382.00	1336.83	376.42	2546.33	405.67
1986	1686.00[b]	527.25	1533.42	545.50	5070.33	533.08	855.25	534.08	1011.00	497.42	2211.58	531.67
1987	1340.33	598.08	1310.25	598.25	5262.50	600.33	1000.17[b]	538.25	1370.33[b]	583.50	1401.50	632.58
1988	843.75	589.17	2029.92[b]	591.33	4562.00	591.83	342.00	566.08	444.08	599.58	846.58	486.08
1989	963.83	606.25	1516.83	619.58	4613.42	617.42	264.92	612.42	786.50	597.50	1645.75	599.42
1990	576.08	782.33	1030.58	882.50	1370.50	811.08	157.17	726.17	119.42	789.25	620.67	777.75
1991	480.83	1067.08[b]	286.67	1064.33	796.58	1068.50	59.50	944.92[b]	34.92[a]	953.3[b]	97.50[a]	756.75
1992	280.58	965.08	299.83	950.67	725.50[a]	950.50	255.75	813.67	118.17	879.92	371.00	643.50
1993	259.17	823.42	515.25	837.92	1266.08	826.83	29.50[a]	683.33	64.33	593.25	135.33	511.67
1994	356.00	890.50	256.42	862.17	765.42	914.25	164.67	892.58	216.00	602.08	350.58	748.17
1995	228.75[a]	1060.67	251.83[a]	1115.42[b]	1153.17	1079.58[b]	321.83	627.33	154.75	704.92	284.33	967.75[b]
Average	835.87	662.02	879.84	670.44	2967.02	669.49	416.27	590.78	621.69	584.72	1311.20	578.56
Range between maximum and minimum	1457.25	687.33	1778.09	756.59	4562.50	695.00	970.67	587.75	1335.41	589.41	2539.75	581.17
S.D.	495.51	241.29	563.62	251.45	1764.43	242.76	276.61	185.33	474.36	177.69	928.13	168.72
C.V.	59.28	36.45	64.06	37.51	59.47	36.26	66.45	31.37	76.30	30.39	70.78	29.16

contd.

Years	MARKETS									
	BEED		MANJLEGAON		LATUR		OSMANABAD		KALLAM	
	Arrivals	*Prices*	*Arrivals*	*Prices*	*Arrivals*	*Prices*	*Arrivals*	*Prices*	*Arrivals*	*Prices*
1981	1169.08	386.67	1483.50	394.83	1017.25	387.50	132.42[b]	383.92	454.42	369.83
1982	1182.08	370.83[a]	1323.42	369.83[a]	443.50	367.75	29.75	363.83	403.92	344.17[a]
1983	1283.08	417.67	1266.67	423.92	1056.92	394.67	75.58	405.83	479.50[b]	394.75
1984	1531.75	429.50	1600.25	421.17	621.75	422.08	30.42	419.67	288.58	406.00
1985	1607.92	383.17	1446.08	371.17	943.08	353.08[a]	28.83	347.17[a]	233.92	370.58
1986	2069.58[b]	515.42	1414.50	528.25	1015.08	490.58	44.42	488.17	289.33	493.25
1987	1204.00	615.00	1683.83[b]	589.17	1029.00	583.33	131.25	603.17	240.83	594.67
1988	792.83	609.17	1025.67	580.42	1147.00[b]	567.92	35.83	548.42	199.75	549.92
1989	850.08	608.75	1274.33	598.42	991.83	577.92	61.58	555.50	179.67	569.25
1990	523.25	799.50	675.42	802.50	690.42	775.00	26.08	783.92	64.75	744.50
1991	33.58[a]	735.92	80.25[a]	1005.75[b]	308.00[a]	1016.25	4.17[a]	402.92	10.92[a]	696.58
1992	201.00	843.00	311.92	892.67	499.50	940.42	6.00	649.67	12.00	917.75
1993	114.67	705.58	244.26	736.25	880.08	777.08	21.00	696.42	22.67	777.00
1994	354.83	709.75	349.92	968.50	938.33	899.17	25.00	800.08	79.75	876.50
1995	267.33	998.50[b]	364.67	984.58	547.92	1030.83[b]	57.25	895.50[b]	71.17	1023.83[b]
Average	878.40	608.56	969.65	644.50	808.64	638.91	47.31	556.28	202.08	615.24
Range between maximum and minimum	2036.00	627.67	1603.58	635.92	839.00	677.75	128.25	548.33	468.58	679.66
S.D.	598.42	184.95	547.26	228.02	255.71	238.21	38.01	170.66	153.91	217.58
C.V.	68.13	30.39	56.44	35.38	31.62	37.28	80.36	30.68	76.16	35.37

a = Minimum level;. b = Maximum level;. S.D. = Standard deviation;. C.V. = Coefficient of variation

Source: As indicated in table 1.

Table 8

Annual Arrivals and Average. Prices of Sunflower at Regulated Markets in Marathwada

(Arrivals in Quintals)
(Price in Rs. per Quintal)

Years	MARKETS											
	PARBHANI		BEED		MANJLEGAON		LATUR		OSMANABAD		KALLAM	
	Arrivals	*Prices*	*Arrivals*	*Prices*	*Arrivals*	*Prices*	*Arrivals*	*Prices*	*Arrivals*	*Prices*	*Arrivals*	*Prices*
1981	172.83	352.25[a]	179.58[a]	319.33[a]	267.92[a]	389.17[a]	4024.83	464.58	651.83	475.00	596.92	446.92
1982	188.42	375.75	379.25	392.50	425.83	405.25	4019.58	450.00	333.50	439.42[a]	715.42	412.00[a]
1983	73.50[a]	414.42	800.17	456.42	516.92	477.00	2977.25	506.67	192.58[a]	494.08	574.25[a]	447.75
1984	130.08	413.08	1347.50	503.75	792.58	490.25	5142.33	528.75	231.92	511.92	974.67	499.75
1985	216.58	415.67	1839.67	440.33	911.33	436.50	9345.58	446.25[a]	215.50	453.17	1422.25	448.33
1986	443.42	523.83	1265.08	591.25	838.33	576.58	1187.75	595.42	229.08	593.58	1228.75	512.42
1987	109.17	599.75	1674.75	695.83	1208.00	736.00	1947.42	755.42	497.67	768.92	1085.83	816.33
1988	115.75	582.92	1903.58	663.33	1774.00	629.75	1173.42[a]	660.83	573.00	659.58	863.08	609.75
1989	115.25	625.42	1439.67	690.00	1471.00	680.33	2776.50	702.50	666.67	718.58	806.33	682.50
1990	543.47	810.00	2278.83	917.42	1892.00	843.50	5257.00	898.75	803.08	894.08	1142.17	859.50
1991	1178.58	1117.92[b]	3797.33[b]	1163.08	2635.00	1116.33	6155.08	1127.50	379.67	1085.67	1274.17	1095.67
1992	2098.50[b]	1058.33	3115.25	1085.58	3966.42[b]	954.33	9209.00	1047.50	558.83	1016.75	801.58	1009.50
1993	1370.75	985.33	3475.83	809.75	2330.75	935.67	13257.50	981.25	1025.50[b]	917.00	1026.83	975.50
1994	954.33	1000.25	1974.33	1150.83	3738.00	1180.17[b]	21562.92[b]	1212.17	498.75	1121.25	1664.92[b]	1145.92
1995	285.58	1005.75	581.92	1242.42[b]	2834.83	1167.08	16802.92	1304.58[b]	636.83	1247.50[b]	964.58	1287.17[b]
Average	533.08	685.38	1736.85	741.45	1706.86	734.53	6989.27	778.81	499.63	759.77	1009.45	749.93
Range between maximum and minimum	2025.00	765.67	3617.75	923.09	3698.50	791.00	20389.50	858.33	832.92	808.08	1090.67	875.17
S.D.	581.18	271.80	1048.93	295.76	1138.66	272.46	5830.55	285.35	232.27	262.39	294.06	286.49
C.V.	109.02	39.66	60.39	39.89	66.71	37.09	83.42	36.64	46.49	34.54	29.13	38.20

a = Minimum level;. b = Maximum level;. S.D. = Standard deviation;. C.V. = Coefficient of variation
Source: As indicated in table 1.

But both standard deviation and coefficient of variation in annual prices of safflower are found to be the highest in case of Lasur (i.e. 251.45 and 37.51 respectively) market and the lowest for Manwat (i.e. 168.72 and 29.16 respectively) market. Arrivals of safflower have exhibited larger year to year variations at Osmanabad market due to greater variations in acreage as well as yield of this crop in the surrounding area of this market. Prices of this commodity have varied to a larger extent in Lasur market due only to low production in this market area.

Sunflower

Sunflower is the most important oilseed crop which is grown in all the seasons in the Marathwada region. In case of sunflower the range between maximum and minimum annual arrivals is found to be highest in case of Latur (20389.50 quintals) market and the lowest in case of Osmanabad (832.92 quintals) market. In case of annual prices of this commodity the range is found to be the highest in case of Beed (923.09 quintals) market and the lowest in case of Parbhani (765.67 quintals) market. The standard deviation and coefficient of variation in arrivals of sunflower are found to be the highest for Latur (5830.55) and Parbhani (109.02) markets respectively and the lowest for Osmanabad (232.27) and Kallam (29.13) markets respectively. The sunflower is grown in all the seasons in the surrounding areas of Latur and Parbhani markets and therefore the arrival of this commodity showed greater ups and downs in these markets. But the interesting thing to note here is that the standard deviation and coefficient of variation in annual prices of sunflower are found to be the highest in case of Beed (295.76 and 39.89 respectively) market and the lowest in case of Osmanabad (262.39 and 34.54 respectively) market. From this it can be said that the annual prices of sunflower have been subject to lesser variation in Osmanabad market during the period of this study. And the prices have varied to a larger extent in Beed market because area and productivity of sunflower are low in this market area.

Groundnut

Groundnut is the most important oilseed crop of the Marathwada region. This crop is grown in two seasons viz., Kharif and summer. The results presented in table 9 about the groundnut commodity

Table 9

Annual Arrivals and Average. Prices of Groundnut at Regulated Markets in Marathwada

(Arrivals in Quintals)
(Price in Rs. per Quintal)

Years	MARKETS									
	JALNA		PARBHANI		MANWAT		BEED		MANJLEGAON	
	Arrivals	Prices	Arrivals	Prices	Arrivals	Prices	Arrivals	Prices	Arrivals	Prices
1981	429.33	352.17[a]	238.83	304.83[a]	154.33	352.25	1523.42	433.75	571.75	409.33
1982	560.92	382.92	181.08	391.33	113.42[a]	341.17[a]	1085.00	409.08	797.42	396.17[a]
1983	970.25	389.75	59.75	473.58	260.75	402.83	780.50	487.83	525.92[a]	477.17
1984	1137.00[b]	467.25	492.83	443.25	633.67	471.17	1908.92	440.00	1229.25	438.58
1985	709.75	389.08	265.92	382.50	1254.33	452.33	1385.75	357.07[a]	978.83	445.83
1986	146.83	542.92	90.17	536.25	211.92	574.92	956.92	595.25	586.75	567.58
1987	47.00[a]	496.33	16.92[a]	505.25	127.33	655.33	900.25	728.33	1026.08	708.50
1988	100.67	613.92	95.17	345.83	167.58	461.50	1457.00	655.00	908.33	631.83
1989	495.17	555.08	1120.67	489.59	862.75	577.92	2966.83[b]	646.25	2446.33	589.75
1990	557.25	612.83	874.58	858.33	235.83	794.75	2286.33	871.17	2489.92	760.25
1991	430.50	981.08[b]	2362.08[b]	1077.08[b]	923.93	928.58	2520.83	1025.83	4332.00[b]	1022.58
1992	183.83	793.83	951.92	1052.42	744.17	991.08	1740.17	1069.67	3409.75	985.25
1993	187.67	853.00	231.67	955.61	135.92	861.83	1126.00	975.00	2183.17	896.50
1994	95.83	809.83	1126.00	930.92	1237.25	739.25	1816.25	1167.08	4124.42	1091.75
1995	111.58	773.50	914.75	769.33	2838.17[b]	1162.86[b]	505.25[a]	1370.75[b]	3910.33	1298.33[b]
Average	410.91	600.90	601.49	634.41	660.09	651.18	1530.63	748.80	1968.02	714.63
Range between maximum and minimum	1090.00	628.51	2345.16	772.25	2724.75	821.69	2461.58	1013.68	3806.08	902.16
S.D.	323.50	191.86	614.40	265.11	704.50	244.04	663.94	303.59	1353.49	274.44
C.V.	78.73	31.93	102.15	41.79	106.73	37.48	43.38	40.54	68.77	38.40

contd.

Years	MARKETS									
	LATUR		UDGIR		OSMANABAD		KALLAM		NANDED	
	Arrivals	*Prices*	*Arrivals*	*Prices*	*Arrivals*	*Prices*	*Arrivals*	*Prices*	*Arrivals*	*Prices*
1981	4509.08	426.50[a]	50.58	455.50	259.92	446.42	242.17	428.67	2265.58	423.42
1982	3434.83	431.42	170.25[b]	431.83	172.25	412.50[a]	323.50	407.25	1155.50	442.75
1983	3573.92	495.00	93.92	436.75	140.33	480.33	139.17	432.58	280.33	488.25
1984	6426.58[b]	498.92	48.08	487.50	389.50	449.00	442.92	383.58[a]	5238.00	381.00[a]
1985	6083.83	506.75	63.50	418.75	209.92	430.58	187.33	461.28	661.75	460.50
1986	5521.25	613.33	21.42	363.00	232.17	585.42	149.50	597.58	107.08[a]	496.25
1987	3746.08	731.67	12.58[a]	547.25	288.17	712.92	100.42[a]	762.25	276.08	588.33
1988	4239.17	688.33	42.17	484.83	309.67	679.83	128.25	562.58	2520.00	492.58
1989	3249.17	620.83	41.67	352.17[a]	491.92[b]	589.25	378.67	559.92	2531.42	652.67
1990	4582.17	866.67	32.25	698.08	413.75	892.00	326.17	779.08	1078.67	737.92
1991	4227.17	1051.67	69.25	1038.08	70.42[a]	915.83	544.92[b]	1010.50	1393.50	1093.33
1992	2322.25	1070.00	112.92	970.58	152.58	991.17	126.33	1007.08	4371.92	1106.17
1993	2515.67	1012.50	99.00	867.17	97.83	909.33	212.83	912.58	5877.67[b]	990.08
1994	4068.75	1128.83	148.3	1207.50[b]	118.50	1080.42	342.42	1131.00	5088.92	1184.92
1995	2245.08[a]	1430.83[b]	57.17	997.25	193.42	1279.92[b]	170.25	1369.67[b]	2389.83	1391.75[b]
Average	4049.67	771.55	70.87	650.42	235.99	723.66	254.32	720.37	2349.08	728.66
Range between maximum and minimum	4181.5	1004.33	157.67	855.33	421.19	867.42	444.50	986.09	5770.59	1010.75
S.D.	1223.29	295.40	44.09	277.36	118.94	263.50	128.28	295.63	1877.39	357.97
C.V.	30.21	38.29	62.20	42.64	50.40	36.41	50.44	41.04	79.92	49.13

a = Minimum level;. b = Maximum level;. S.D. = Standard deviation;. C.V. = Coefficient of variation

Source: As indicated in table 1.

reveal that the range between maximum and minimum annual arrivals of this commodity is the highest in case of Nanded (5790.59 quintals) market and the lowest in case of Udgir (157.67 quintals) market. The range between the maximum and minimum annual prices of groundnut is found to be the highest in case of Beed (Rs. 1073.68) market and the lowest in case of Jalna (Rs. 628.91) market. The standard deviation and coefficient of variation in arrivals of groundnut are found to be the highest in case of Nanded (1877.39) and Manwat (106.73) markets respectively and the lowest for Udgir (44.09) and Latur (30.21) markets respectively. It means that the degree of annual variation in arrivals of groundnut has been relatively higher in Nanded and Manwat markets and relatively smaller in Udgir and Latur markets during the period of this study. The larger annual variation in arrivals of groundnut at Nanded and Manwat markets is due to year to year variations in yield and acreage of this crop in these market areas. In these market areas, the groundnut crop is mostly grown in summer season on irrigated lands. The availability of canal irrigation water for this crop has varied from year to year which explains the larger variations in arrivals of this commodity in Nanded and Manwat markets. In case of annual prices of groundnut both standard deviation and coefficient of variation are found to be the highest in case of Nanded (357.97 and 49.13 respectively) market and the lowest in case of Jalna (191.86 and 31.93 respectively) market.

The present study has indicated that the annual variations in arrivals of wheat were higher in Latur and Udgir markets and lower in Osmanabad and Manjlegaon markets. But the prices of wheat showed a higher degree of variation in Partur market and lower in Naigaon market during the study period.

In case of bajra, the study indicated that the annual variations in arrivals were relatively higher in Kallam market and relatively lower in Aurangabad market. Annual prices of bajra showed highest variation in Lasur market and lowest in Udgir market. Arrivals of kharif jowar have exhibited the highest degree of annual variation in Lasur market and lowest in Latur market. This is indicated by the coefficient of variation in arrivals for these markets. Prices of kharif jowar have varied to a larger extent in Beed and Aurangabad markets and to a smaller extent in Partur market.

This study has indicated that the annual arrivals of rabi jowar have been subject to larger variation in Jalna and Udgir markets and smaller variation in Kallam market during the study period. But the annual variation in its prices is found to be the highest in Lasur market and the lowest in Nanded and Parbhani markets.

In this study it is found that the annual variation in arrivals of tur was the highest in Latur and Aurangabad markets and lowest in Naigaon and Parbhani markets during the period of this study. But the degree of annual variations in its prices was the highest in Latur and Lasur markets and lowest in Naigaon market. The degree of year to year variations in arrivals of greengram was the highest for Jalna and Aurangabad markets and the lowest for Kallam and Partur markets. But the annual prices of this commodity have exhibited larger variations in Udgir market and smaller variations in Parbhani market.

In case of safflower, annual variation in arrivals was the highest in case of Osmanabad market and lowest in Latur market. But the degree of variation in its annual prices was the highest in case of Lasur and lowest for Manwat market during the study period.

This study has indicated that the annual variation of arrivals of sunflower was the highest in Latur and Parbhani markets and lowest in Osmanabad and Kallam markets. But the annual price variation of this commodity was the highest in Beed market and the lowest in Osmanabad market.

In this study it is found that the degree of variation in annual arrivals of groundnut was the highest in Nanded and Manwat markets and lowest in Udgir and Latur markets. But prices of this commodity have varied to a greater extent in Nanded market and to a smaller extent in Jalna market during the study period.

5

ANALYSIS OF TRENDS IN ARRIVALS AND PRICES OF AGRICULTURAL COMMODITIES

It is the well known fact that efficient marketing plays an important role not only in raising economic activity but also in making the most economic use of what is produced, which is an important determinant of the level and the rate of production, depends also on the efficiency of marketing. Maximum production is essential but its prudent marketing is equally important. Market imperfections contribute to higher price spread and wastage, while inefficient production techniques push up the cost of production.

The pricing system should ensure fair returns to all the parties involved in the marketing process and it should provide an adequate incentive to further production. But in India, due to unorganised nature of agricultural markets, the pricing system is often biased and exploitative. On account of many factors, agricultural markets did not encourage competitive prices to farmers to guide their production programmes. To remove these defects and to ensure fair prices to farmers, the Government of India took series of measures like regulation of markets, price support programmes, constitution of agricultural prices commission, etc.

Prices determine not only what shall be produced but also how much should be produced and their distribution. For achieving the targets of agricultural production, it is important that growers should have the full confidence that additional efforts and investments will yield adequate returns. The wide fluctuations in the prices affect the growers capacity of making sustained efforts for increasing production. These fluctuations in the prices of agricultural products are the greatest hurdle in the way of agricultural development.

The analysis of the movements in prices helps in ascertaining as to what extent the marketing system is efficient and also helps in production planning. Further, the behaviour of prices over a period of time helps to understand the price structure and movements. It is necessary to study the farm prices which forms an essential requirement for policy formulation. On the basis of price studies the cultivators can make decisions for proper allocation of crop areas by anticipating future prices based on the prices which prevented in the past. The knowledge of structure and movement of prices and arrivals and of their relationship is essential for enabling the farmers to reap the full benefit of their harvests.

Trend is the basic tendency of a series to grow or decline over a period of time. It measures the changes either up or down, of the values over a long period of time. The trend is that component of variation reveals general direction of change over a period of time. Trend is usually established on the basis of at least 10-15 years data. The concept of trend does not include short range oscillations but rather steady movements over a long time. Under trend, changes (increases or decreases) are not also sudden but are slow moving tendencies. This component in price variation is affected by such factors as changes in demand arising from changes in population, income, dietary habits, customs and establishment of processing industries for the commodity concerned. Price trend is also affected by adjustment in supply arising out of development of cold storage and marketing facilities, production technology and market arrivals. Thus, the long term smooth movements are the results of the forces which change gradually over a long period of time. As a result, a number of factors including the changes in prices the variation in arrivals is caused. Further, the arrivals trend is affected by the changes in production and prices.

In this chapter an attempt is made to study the trend in arrivals and prices of main foodgrains and oilseeds at regulated markets in Marathwada.

The trend in arrivals of an agricultural commodity is studied by fitting the linear regression equation of arrivals on time with the help of least- squares method. Similarly, the trend in prices of an agricultural commodity is studied by fitting the linear regression equation of prices on time with the help of least-squares method. The following linear regression equation used to estimate the annual trend values of arrivals and prices of each agricultural commodity for each regulated market.

$$Y = a + bx$$

where, Y = Arrival / Price

x = time (in years)

a = intercept, and

b = slope.

A time series data on annual arrivals and prices of selected agricultural commodities used in the present analysis pertains to the period from 1981 to 1995. The slope and the intercept values of the estimated regression equations are tested for their significance by using the 't' test.

This chapter is divided into two sections. Section 'A' presents the analysis of trend in arrivals and prices of foodgrains and Section 'B' present the analysis of the trend in annual arrivals and prices of oilseeds.

SECTION A : TRENDS IN ARRIVALS AND PRICES OF FOODGRAINS

This section deals with the trend in annual arrivals and prices of foodgrains, namely, wheat, bajra, kharif jowar, rabi jowar, tur and greengram at the regulated markets in Marathwada.

Wheat

The annual trend values of arrivals and prices of wheat for regulated markets are given in table 1. It is observed from this table that annual trend values of arrivals of wheat have shown rising trend over the period in seven regulated markets viz., Aurangabad, Lasur, Jalna, Osmanabad, Kallam, Nanded and Manjlegaon. Among the markets where annual trend values of arrivals showed rising trend, the highest base year value of arrivals (i.e. intercept) is found in case of Jalna

Table 1
Trend Values of Arrivals and Prices of Wheat at Regulated Markets in Marathwada

(Arrivals in '00 Quintals)
(Price in Rs. per Quintal

YEARS	AURANGABAD		LASUR		JALNA		MARKETS PARTUR		LATUR		UDGIR		OSMANABAD	
	Trend values of arrivals	Trend values of prices	Trend values of arrivals	Trend values of prices	Trend values of arrivals	Trend values of prices	Trend values of arrivals	Trend values of prices	Trend values of arrivals	Trend values of prices	Trend values of arrivals	Trend values of prices	Trend values of arrivals	Trend values of prices
1981	191.09	174.26	67.29	166.15	413.80	171.10	69.80	142.52	398.68	164.11	83.74	171.74	23.56	173.96
1982	210.18	196.42	84.39	189.99	421.72	195.81	67.49	170.62	379.69	190.82	77.19	193.95	24.46	200.52
1983	229.27	218.58	101.49	213.83	429.56	220.52	65.18	198.72	360.70	217.53	70.64	216.16	25.36	227.08
1984	248.36	240.74	118.59	237.67	437.40	245.23	62.87	226.82	341.71	244.24	64.09	238.37	26.26	253.64
1985	267.45	252.90	135.69	261.51	445.24	269.94	60.56	254.92	322.72	270.95	57.54	260.58	27.16	280.20
1986	286.54	285.06	152.79	285.35	453.08	294.65	58.25	283.02	303.73	297.66	50.99	282.79	28.06	306.76
1987	305.63	307.22	169.89	309.19	460.92	319.36	55.94	311.12	284.74	324.37	44.44	305.00	28.96	333.32
1988	324.72	329.38	186.99	333.03	468.76	344.07	53.63	339.22	265.75	351.08	37.89	327.21	29.86	359.88
1989	343.81	351.54	204.09	356.87	476.60	368.78	51.32	367.32	246.76	377.79	31.34	349.42	30.76	386.44
1990	362.90	373.70	221.19	380.71	484.44	393.49	49.01	395.42	227.77	404.50	24.79	371.63	31.66	413.00
1991	381.99	395.86	238.29	404.55	492.28	418.20	46.70	423.52	208.78	431.21	18.24	393.84	32.56	439.56
1992	401.08	418.02	255.39	428.39	500.12	442.91	44.39	451.62	189.79	457.92	11.69	416.05	33.46	466.12
1993	420.17	440.18	272.49	452.23	507.96	467.62	42.08	479.72	170.80	484.63	5.14	438.26	34.36	492.68
1994	439.26	462.34	289.59	476.07	515.80	492.33	39.77	507.82	151.81	511.34	-1.41	460.47	35.26	519.24
1995	458.35	484.50	306.69	499.91	523.64	517.04	37.46	535.92	132.82	538.05	-7.96	482.68	36.16	545.80

contd.

YEARS	KALLAM		PARBHANI		MANWAT		MARKETS NANDED		NAIGAON		BEED		MANJLEGAON	
	Trend values of arrivals	Trend values of prices	Trend values of arrivals	Trend values of prices	Trend values of arrivals	Trend values of prices	Trend values of arrivals	Trend values of prices	Trend values of arrivals	Trend values of prices	Trend values of arrivals	Trend values of prices	Trend values of arrivals	Trend values of prices
1981	25.74	163.29	127.95	180.37	77.54	159.71	172.25	164.31	65.56	157.31	164.16	164.29	54.35	168.93
1982	28.93	189.05	119.36	203.17	74.86	185.65	176.88	186.88	61.75	174.80	163.67	191.91	54.54	194.02
1983	32.12	214.81	110.77	225.97	72.18	211.59	181.51	209.45	57.94	192.29	163.18	219.53	54.73	219.11
1984	35.31	240.57	102.18	248.77	69.50	237.53	186.14	232.02	54.13	209.78	162.69	247.15	54.92	244.20
1985	38.50	266.33	93.59	271.57	66.82	263.47	190.77	254.59	50.32	227.27	162.20	274.77	55.11	269.29
1986	41.69	292.09	85.00	294.37	64.14	289.41	195.40	277.16	46.51	244.76	161.71	302.39	55.30	294.38
1987	44.88	317.85	76.41	317.17	61.46	315.35	200.03	299.73	42.70	262.25	161.22	330.01	55.49	319.47
1988	48.07	343.61	67.82	339.97	58.78	341.29	204.66	322.30	38.89	279.74	160.73	357.63	55.68	344.56
1989	51.26	369.37	59.23	362.77	56.10	367.23	209.29	344.87	35.08	297.23	160.24	385.25	55.87	369.65
1990	54.45	395.13	50.64	385.57	53.42	393.17	213.92	367.44	31.27	314.72	159.75	412.87	56.06	394.74
1991	57.64	420.89	42.05	408.37	50.74	419.11	218.55	390.01	27.46	332.21	159.26	440.49	56.25	419.83
1992	60.83	446.65	33.46	431.17	48.06	445.05	223.18	412.58	23.65	349.70	158.77	468.11	56.44	444.92
1993	64.02	472.41	24.87	453.97	45.38	470.99	227.81	435.15	19.84	367.19	158.28	495.73	56.63	470.01
1994	67.21	498.17	16.28	476.77	42.70	496.93	232.44	457.72	16.03	384.68	157.79	523.35	56.82	495.10
1995	70.40	523.93	7.69	499.57	40.02	522.87	237.07	480.29	12.22	402.17	157.30	550.97	57.01	520.19

Source: Data on yearly arrivals and prices are compiled from the office records of the regulated markets.

Table 2
The Linear Trend Equations for Arrivals and Prices of Wheat in Regulated Markets in Marathwada

Markets	*Arrivals*		*Prices*	
	a	*b*	*A*	*b*
Aurangabad	172.00^{NS}	19.09**	152.10***	22.16***
	(165.16)	(9.87)	(43.45)	(2.60)
Lasur	50.19^{NS}	17.10***	142.31***	3.84***
	(84.24)	(5.03)	(38.48)	(2.30)
Jalna	406.04**	7.84^{NS}	146.39***	4.71***
	(194.78)	(11.64)	(42.54)	(2.54)
Partur	72.11***	-2.31*	114.42**	28.10***
	(26.31)	(1.57)	(57.48)	(3.83)
Parbhani	136.54***	-8.59***	157.57***	22.80***
	(28.18)	(1.68)	(43.18)	(2.58)
Manwat	80.22**	-2.68^{NS}	133.77***	25.94***
	(40.46)	(2.42)	(49.25)	(2.94)
Beed	164.65**	-0.49^{NS}	136.67***	27.62***
	(71.89)	(4.30)	(49.27)	(2.94)
Manjlegaon	54.16**	0.19^{NS}	143.84***	25.09***
	(22.38)	(1.34)	(44.60)	(2.65)
Latur	417.67**	-18.99**	137.40***	26.71***
	(208.04)	(12.43)	(38.94)	(3.52)
Udgir	90.29**	-6.55***	149.53***	22.21***
	(38.75)	(2.32)	(47.60)	(2.84)
Osmanabad	22.6^{NS}	0.90^{NS}	147.40***	26.26***
	(20.22)	(1.21)	(47.73)	(2.85)
Kallam	22.55^{NS}	3.19**	137.53**	25.76***
	(29.21)	(1.75)	(60.03)	(3.59)
Nánded	167.62*	4.63^{NS}	141.74***	22.57***
	(107.57)	(6.43)	(41.48)	(2.48)
Naigaon	69.37**	-3.81**	39.82^{NS}	17.49***
	(29.78)	(1.78)	(44.01)	(2.63)

Note: Bracketed figures are standard errors.
a = Intercept, b = Slope,
*** Significant at 1 per cent level, ** Significant at 5 per cent level,
* Significant at 10 per cent level, NS = Not-significant.

market (40604 quintals) and lowest in case of Kallam market (2255 quintals). The annual rate of increase in arrivals of wheat is found to be the highest in case of Aurangabad market (1909 quintals) and the lowest in case of Osmanabad (90 quintals) market during the period from 1981 to 1995. It means that the arrivals of wheat have increased at the highest annual rate in Aurangabad market and at the lowest rate in Osmanabad market during the study period. The wheat crop arrivals have enhanced on account of increase in area under cultivation in these market areas. The annual rate of increase in arrivals of wheat is found to be statistically significant in case of Lasur (at one per cent level) and in case of Aurangabad and Kallam (at 5 per cent level) markets. The annual rate of increase is not found statistically significant in case of Jalna, Osmanabad, Manjlegaon and Nanded markets. Among the markets in seven markets annual trend values of arrivals of wheat have shown a decreasing trend during the study period. These markets are: Partur, Latur, Udgir, Parbhani, Manwat, Naigaon and Beed. The highest base year value of arrivals of wheat (i.e. intercept) is found in case of Latur (41767 quintals) and the lowest in case of Naigaon (6937 quintals) market. The annual rate of decrease in arrivals of wheat is found to be the highest in case of Latur (-1899 quintals) and the lowest in case of Beed (-49 quintals) during the study period. In these market areas, the average yield per hectare of wheat has decreased and further that the farmers shifted emphasis from wheat to other cash crops. The annual rate of decrease in arrivals of wheat is found to be statistically significant at 1 per cent level in case of Udgir and Parbhani, at 5 per cent level in case of Latur and Naigaon markets, at 10 per cent level in case of Partur, and statistically insignificant in case of Beed and Manwat markets.

The annual trend values of prices of wheat have shown an increasing trend in all the markets during the study period. The highest base year value of price (i.e. intercept) is found in case of Parbhani (Rs. 157.57) and the lowest in case of Naigaon (Rs. 39.82) market. The annual rate of increase in the price of wheat is found to be the highest in case of Partur (Rs. 28.10) market and the lowest for Naigaon (Rs. 17.49) market during the period under consideration. Further, the annual rate of increase in wheat price is statistically significant at 1 per cent level in all the markets.

Bajra

Trend values for annual arrivals of bajra in markets are given in table 3. It is observed from the table that annual arrivals of bajra have shown increasing trend over the time in seven markets viz., Aurangabad, Lasur, Jalna, Partur, Latur, Kallam and Beed. Among the markets where annual trend values of arrivals of this commodity

Table 3
Trend Values of Arrivals and Prices of Bajra at Regulated Markets in Marathwada

(Arrivals in '00 Quintals)
(Price in Rs. per Quintal)

YEARS	MARKETS									
	AURANGABAD		LASUR		JALNA		PARTUR		LATUR	
	Trend values of arrivals	Trend values of prices	Trend values of arrivals	Trend values of prices	Trend values of arrivals	Trend values of prices	Trend values of arrivals	Trend values of prices	Trend values of arrivals	Trend values of prices
1981	243.26	103.40	273.65	88.48	515.89	86.51	35.25	69.76	422.77	87.10
1982	255.00	120.90	273.65	110.64	568.46	106.18	36.80	88.23	431.79	105.55
1983	266.74	138.40	273.65	132.80	621.03	125.85	38.35	106.70	440.81	124.00
1984	278.48	155.90	273.65	154.96	673.60	145.52	39.90	125.17	449.83	142.45
1985	290.22	173.40	273.65	177.12	726.17	165.19	41.45	143.64	458.85	160.90
1986	301.96	190.90	273.65	199.28	778.74	184.86	43.00	162.11	467.87	179.35
1987	313.70	208.40	273.65	221.44	831.31	204.53	44.55	180.58	476.89	197.80
1988	325.44	225.90	273.65	243.60	883.88	224.20	46.10	199.05	485.91	216.25
1989	337.18	243.40	273.65	265.76	936.45	243.87	47.65	217.52	494.93	234.70
1990	348.92	260.90	273.65	287.92	989.02	263.54	49.20	235.99	503.95	253.15
1991	360.66	278.40	273.65	310.08	1041.59	283.21	50.75	254.46	512.97	271.60
1992	372.40	295.90	273.65	332.24	1094.16	302.88	52.30	272.93	521.99	290.05
1993	384.14	313.40	273.65	354.40	1146.73	322.55	53.85	291.40	531.01	308.50
1994	395.88	330.90	273.65	376.56	1199.30	342.22	55.40	309.87	540.03	326.95
1995	407.62	348.40	273.65	398.72	1251.87	361.89	56.95	328.34	549.05	345.40

contd.

YEARS	MARKETS									
	UDGIR		KALLAM		MANWAT		BEED		MANJLEGAON	
	Trend values of arrivals	Trend values of prices	Trend values of arrivals	Trend values of prices	Trend values of arrivals	Trend values of prices	Trend values of arrivals	Trend values of prices	Trend values of arrivals	Trend values of prices
1981	59.30	96.99	-2.43	50.38	31.50	81.22	312.32	97.28	204.48	87.87
1982	55.78	111.49	0.49	71.06	31.07	95.99	349.09	117.55	203.98	108.09
1983	52.26	125.99	3.41	91.74	30.64	110.76	385.86	137.82	203.48	128.31
1984	48.74	140.49	6.33	112.42	30.21	125.53	422.63	158.09	202.98	148.53
1985	45.22	154.99	9.25	133.10	29.78	140.30	459.40	178.36	202.48	168.75
1986	41.70	169.49	12.17	153.78	29.35	155.07	496.17	198.63	201.98	188.97
1987	38.18	183.99	15.09	174.46	28.92	169.84	532.94	218.90	201.48	209.19
1988	34.66	198.49	18.01	195.14	28.49	184.61	569.71	239.17	200.98	229.41
1989	31.14	212.99	20.93	215.82	28.06	199.38	606.48	259.44	200.48	249.63
1990	27.62	227.49	23.85	236.50	27.63	214.15	643.25	279.71	199.98	269.85
1991	24.10	241.99	26.77	257.18	27.20	228.92	680.02	299.98	199.48	290.07
1992	20.58	256.49	29.69	277.86	26.77	243.69	716.79	320.25	198.98	310.29
1993	17.06	270.99	32.61	298.54	26.34	258.46	753.56	340.52	198.48	330.51
1994	13.54	285.49	35.53	319.22	25.91	273.23	790.33	360.79	197.98	350.73
1995	10.02	299.99	38.45	339.90	25.48	288.00	827.10	381.06	197.48	370.95

Source: As indicated in table 1.

showed rising trend, the highest base year value of arrivals (i.e. intercept) is found in case of Jalna (46332 quintals) market and lowest in case of Kallam (-535 quintals) market. The annual rate of increase in arrivals of bajra is found to be the highest in case of Jalna (5257 quintals) market and the trend values of arrivals showed no change in case of Lasur market. It means that the arrivals of bajra have increased at the highest annual rate in Jalna market. But the trend value of arrival of this commodity remained constant in case of Lasur market during the study period. As the bajra crop requires less rainfall and unfortunately the rains in the recent years are poor, farmers have started growing more bajra. So, the bajra arrivals in these markets have showed an increase. The annual rate of increase in arrivals of bajra is found to be statistically significant in case of Kallam (at one per cent level), Beed (five per cent level) and Aurangabad and Jalna (ten per cent level) markets. But the annual rate of increase of arrivals of this commodity is not found statistically significant in case of Lasur, Partur and Latur markets (see Table 4). Among the selected markets annual trend values of arrivals of bajra have shown a decreasing trend in Manjlegaon, Udgir and Manwat markets during the study period. Among these markets, the highest base year value of arrival of bajra (i.e. intercept) is found in case of Manjlegaon (20498 quintals) and the lowest in case of Manwat (3113 quintals) market. The annual rate of decrease in arrivals of bajra is found to be the highest in case of Udgir (-352 quintals) and the lowest for Manwat market (-43 quintals) during the study period. In the three market areas, the irrigation is satisfactory, which has enabled farmers to shift from bajra to other irrigated high value crops. The annual rate of decrease in arrivals of bajra is found to be statistically significant at 1 per cent level in case of Udgir and statistically insignificant in case of Manwat and Manjlegaon markets.

The annual trend values of prices of bajra have shown an increasing trend in all the selected markets during the study period. The highest base year value of price (i.e. intercept) is found in case of Aurangabad (Rs. 85.90) market and the lowest in case of Kallam (Rs. 29.70) market. The annual rate of increase in the price of bajra is found to be the highest in case of Lasur (Rs. 22.16) market and the lowest for Udgir (Rs. 14.50) market during the study period. Further, the annual rate of increase in bajra price is found statistically significant at 1 per cent level in all the markets.

Table 4
The Linear Trend Equations for Arrivals and Prices of Bajra in Regulated Markets in Marathwada

Markets	*Arrivals*		*Price*	
	a	*b*	*A*	*b*
Aurangabad	231.52**	11.74*	85.90^{NS}	17.50***
	(130.34)	(7.79)	(73.27)	(4.38)
Lasur	273.65*	0.00^{NS}	66.32^{NS}	22.16***
	(189.30)	(11.31)	(70.86)	(4.23)
Jalna	463.32^{NS}	52.57*	66.84^{NS}	19.67***
	(563.43)	(31.67)	(59.87)	(3.58)
Partur	33.70 *	1.55^{NS}	51.24^{NS}	18.47***
	(24.96)	(1.49)	(54.47)	(3.26)
Manwat	31.13*	-0.43^{NS}	66.45*	14.77***
	(19.35)	(1.15)	(37.67)	(2.25)
Beed	275.55^{NS}	36.77**	77.01^{NS}	20.37***
	(242.36)	(14.48)	(66.43)	(3.97)
Manjlegaon	204.98*	-0.50^{NS}	67.65^{NS}	20.22***
	(118.95)	(7.11)	(59.60)	(3.56)
Latur	413.75*	9.02^{NS}	68.65^{NS}	18.45***
	(288.46)	(17.24)	(60.92)	(3.64)
Udgir	62.82***	-3.52***	82.49**	14.50***
	(20.65)	(1.23)	(46.45)	(2.78)
Kallam	-5.35^{NS}	2.92***	29.70^{NS}	20.68***
	(14.26)	(0.85)	(64.63)	(3.86)

Note: Bracketed figures are standard errors.
a = Intercept, b = Slope,
*** Significant at 1 per cent level, ** Significant at 5 per cent level,
* Significant at 10 per cent level, NS = Not-significant.

Rabi Jowar

The annual trend values of arrivals and prices of rabi jowar for regulated markets are given in table 5. It is observed from this table that annual trend values of arrivals of rabi jowar have shown increasing trend over the period of study in four regulated markets. The markets where the trend values of arrivals are increasing include Jalna, Osmanabad, Beed and Nanded. Among these markets the highest base year value of arrivals (i.e. intercept) is found in case of Jalna (68045 quintals) and the lowest in case of Osmanabad (3716 quintals). The annual rate of increase in arrivals of this commodity is found to be the highest in case of Jalna (3432 quintals) market and the lowest in case of Osmanabad (243 quintals) market. It means that the arrivals of rabi jowar have increased at the highest annual

Table 5

Trend Values of Arrivals and Prices of Rabi Jowar at Regulated Markets in Marathwada

(Arrivals in '00 Quintals)
(Price in Rs. per Quintal

YEARS	AURANGABAD		LASUR		JALNA		PARTUR		LATUR		UDGIR		OSMANABAD	
	Trend values of arrivals	Trend values of prices	Trend values of arrivals	Trend values of prices	Trend values of arrivals	Trend values of prices	Trend values of arrivals	Trend values of prices	Trend values of arrivals	Trend values of prices	Trend values of arrivals	Trend values of prices	Trend values of arrivals	Trend values of prices
1981	275.31	131.87	425.35	101.06	714.77	124.71	204.38	110.14	265.32	127.43	40.36	132.43	39.59	137.08
1982	272.27	157.74	421.01	131.82	749.09	150.99	194.36	136.17	255.65	156.49	36.83	157.24	42.02	162.74
1983	269.23	183.61	416.67	162.58	783.41	177.27	184.34	162.20	245.98	185.55	33.30	182.05	44.45	188.40
1984	266.99	209.48	412.23	193.34	817.73	203.55	174.32	188.23	236.11	214.61	29.77	206.86	46.88	214.06
1985	263.15	235.35	407.99	224.10	852.05	229.83	164.30	214.26	226.64	243.67	26.44	231.67	49.31	239.72
1986	260.11	261.22	403.65	254.86	886.37	256.11	154.28	240.29	216.97	272.73	22.71	256.48	51.74	265.38
1987	257.07	287.09	399.31	285.62	920.69	282.39	144.26	266.32	207.30	301.79	19.18	281.29	54.17	291.04
1988	254.03	312.96	394.97	316.38	955.01	308.67	134.24	292.35	197.63	330.85	15.65	306.10	56.60	316.70
1989	250.99	338.83	390.63	347.14	989.33	334.95	124.22	318.38	187.96	359.91	12.12	330.91	59.03	342.36
1990	247.95	364.70	386.29	377.90	1023.65	361.23	114.20	344.41	178.29	388.97	8.59	355.72	61.46	368.02
1991	244.91	390.57	381.95	408.66	1057.97	387.51	104.18	370.44	168.62	418.03	5.06	380.53	63.89	393.68
1992	241.87	416.44	377.61	439.42	1092.29	413.79	94.16	396.47	158.95	447.09	1.53	405.34	66.32	419.34
1993	238.83	442.31	373.27	470.18	1126.61	440.07	84.14	422.50	149.28	476.15	-2.00	430.15	68.75	445.00
1994	235.79	468.18	368.93	500.94	1160.93	466.35	74.12	448.53	139.61	405.21	-5.53	454.96	71.18	470.66
1995	232.75	494.05	364.59	531.70	1195.25	492.63	64.10	474.56	129.94	534.27	-9.06	479.77	73.61	496.32

contd.

YEAR	MARKETS											
	KALLAM		PARBHANI		MANWAT		NANDED		BEED		MANJLEGAON	
	Trend values of arrivals	Trend values of prices	Trend values of arrivals	Trend values of prices	Trend values of arrivals	Trend values of prices	Trend values of arrivals	Trend values of prices	Trend values of arrivals	Trend values of prices	Trend values of arrivals	Trend values of prices
1981	64.08	132.80	165.06	143.29	211.36	114.28	138.26	104.64	399.59	115.32	528.50	112.18
1982	62.83	158.71	153.44	166.95	201.27	140.53	156.40	119.92	408.43	142.87	502.84	139.33
1983	61.58	184.62	141.82	190.61	191.81	166.78	174.54	135.20	417.27	170.42	476.88	166.48
1984	60.33	210.53	130.20	214.27	181.09	193.03	192.68	150.48	426.11	197.97	450.92	193.63
1985	59.08	236.44	118.58	237.93	171.00	219.28	210.82	165.76	434.95	225.52	424.96	220.78
1986	57.83	262.35	106.96	261.59	160.91	245.53	228.96	181.04	443.79	253.07	399.00	247.93
1987	56.58	288.26	95.34	285.25	150.82	271.78	247.10	196.32	452.63	280.62	373.04	275.08
1988	55.33	314.17	83.72	308.91	140.73	298.03	265.24	211.60	461.47	308.17	347.08	302.23
1989	54.08	340.08	72.10	332.57	130.64	324.28	283.38	226.88	470.31	335.72	321.12	329.38
1990	52.83	365.99	60.48	356.23	120.55	350.53	301.52	242.16	479.15	363.27	295.16	356.53
1991	51.58	391.90	48.86	379.89	110.46	376.78	319.66	257.44	487.99	390.82	269.20	383.68
1992	50.33	417.81	37.24	403.55	100.37	403.03	337.80	272.72	496.83	418.37	243.24	410.83
1993	49.08	443.72	25.62	427.21	90.28	429.28	355.94	288.00	505.67	445.92	217.28	437.83
1994	47.83	469.63	14.00	450.87	80.19	455.53	374.08	303.28	514.51	473.47	191.32	465.13
1995	46.58	495.54	2.38	474.53	70.10	481.78	392.22	318.56	523.35	501.02	165.36	492.28

Source: As indicated in table 1.

Table 6
The Linear Trend Equations for Arrivals and Prices of Rabi Jowar in Regulated Markets In Marathwada

Markets	*Arrivals*		*Price*	
	a	*b*	*A*	*b*
Aurangabad	-278.35***	-3.04NS	106.00*	25.87***
	(94.25)	(5.63)	(74.09)	(4.43)
Lasur	429.69***	-4.34NS	70.30NS	30.76***
	(156.81)	(9.37)	(82.16)	(4.91)
Jalna	680.45**	34.32*	98.43*	26.28***
	(332.56)	(19.87)	(72.63)	(4.34)
Partur	214.40***	-10.02***	84.11NS	26.03***
	(49.79)	(2.98)	(75.13)	(4.49)
Parbhani	176.68***	-11.62***	119.63**	23.66***
	(36.80)	(2.20)	(58.77)	(3.51)
Manwat	221.45***	-10.09***	88.03NS	26.25***
	(58.59)	(3.50)	(74.01)	(4.42)
Beed	390.75**	8.84NS	87.77NS	27.55***
	(153.93)	(9.20)	(80.84)	(4.83)
Manjlegaon	554.76**	-25.96***	85.03NS	27.15***
	(153.89)	(9.20)	(72.85)	(4.35)
Latur	274.94**	-9.67NS	98.37NS	29.06***
	(129.23)	(7.72)	(76.88)	(4.59)
Udgir	43.89**	-3.53**	107.62NS	24.81***
	(19.75)	(1.18)	(90.66)	(5.42)
Osmanabad	37.16*	2.43**	111.42*	25.66***
	(22.52)	(1.35)	(74.58)	(4.46)
Kallam	65.33***	-1.25NS	106.89*	25.91***
	(17.99)	(1.08)	(76.47)	(4.57)
Nanded	120.12NS	18.14**	89.36*	15.28***
	(151.39)	(9.05)	(62.65)	(3.74)

Note: Bracketed figures are standard errors.
a = Intercep t, b = Slope,
*** Significant at 1 per cent level, ** Significant at 5 per cent level,
* Significant at 10 per cent level, NS = Not-significant.

rate in Jalna market and at the lowest rate in Osmanabad market during the study period. These markets showed increased arrivals of the crop which may be attributed to increase in production and the farmers expectations for better prices. The annual rate of increase in arrivals of rabi jowar is found to be statistically significant in case of Osmanabad, Nanded (at 5 per cent level) and Jalna (at 10 per cent level) markets. The annual rate of increase in arrivals is not found statistically significant in case of Beed market. Among the markets annual trend values of arrivals of rabi jowar have shown a decreasing trend during the study period. These markets are: Aurangabad, Lasur, Partur, Latur, Udgir, Parbhani, Manwat, Kallam and Manjlegaon. Among these markets the highest base year value of arrivals of rabi jowar (i.e. intercept) is found in case of Manjlegaon (55476 quintals) market and the lowest in case of Aurangabad (-278.35 quintals) market. The annual rate of decrease in arrivals of rabi jowar is found to be the highest in case of Manjlegaon (-2596 quintals) market and the lowest in case of Kallam (-125 quintals) market during the study period. The decreasing trend is observed due to the reduced production of the crop and poor weather conditions during the period under study. The annual rate of decrease in arrivals of rabi jowar is found statistically significant at 1 per cent level in case of Partur, Manwat, Manjlegaon and Parbhani markets, at 5 per cent level in case of Udgir and statistically insignificant in case of Aurangabad, Lasur, Latur and Kallam markets.

The annual trend values of prices of rabi jowar have shown an increasing trend in all the markets during the study period. The highest base year value of price of rabi jowar (i.e. intercept) is found in case of Parbhani (Rs. 119.63) and the lowest in case of Lasur (Rs. 70.30) market. The annual rate of increase in the prices of rabi jowar is found to be the highest in case of Lasur (Rs. 30.76) market and the lowest for Nanded (Rs. 15.28) market during the study period. Further, the annual rate of increase in rabi jowar prices is statistically significant at 1 per cent level in all the markets.

Kharif Jowar

The annual trend values of arrivals and prices of kharif jowar for regulated markets are given in table 7. It is observed from this table that annual trend values of arrivals of kharif jowar have shown increasing trend over the period of this study in two regulated markets viz., Latur and Naigaon. The markets where annual trend values of arrivals showed increasing trend, the highest base year value of arrivals (i.e. intercept) is found in case of Latur (178629 quintals) market and the lowest in case of Naigaon (10854 quintals) market. The annual rate of increase is found to be the highest in case of Latur (1095 quintal) market and the lowest in case of Naigaon (15 quintals) market during the study period. It means that the arrivals of kharif

Table 7
Trend Values of Arrivals and Prices of Kharif Jowar at Regulated Markets in Marathwada

(Arrivals in '00 Quintals)
(Price in Rs. per Quintal)

YEARS	MARKETS											
	AURANGABAD		LASUR		JALNA		PARTUR		LATUR		UDGIR	
	Trend values of arrivals	Trend values of prices	Trend values of arrivals	Trend values of prices	Trend values of arrivals	Trend values of prices	Trend values of arrivals	Trend values of prices	Trend values of arrivals	Trend values of prices	Trend values of arrivals	Trend values of prices
1981	387.23	90.04	199.14	91.30	827.20	88.56	70.16	85.70	1797.24	79.84	601.88	79.33
1982	372.44	104.84	185.97	104.69	798.95	104.33	68.70	98.38	1808.19	96.92	598.87	93.60
1983	357.65	119.64	172.80	118.08	770.70	120.10	67.24	111.06	1819.14	114.00	595.86	107.87
1984	342.86	134.44	159.63	131.47	742.45	135.87	65.78	123.74	1830.09	131.08	592.85	122.14
1985	328.07	149.24	146.46	144.86	714.20	151.64	64.32	136.42	1841.04	148.16	589.84	136.41
1986	313.28	164.04	133.29	158.25	685.95	167.41	62.86	149.10	1851.99	165.24	586.83	150.68
1987	298.49	178.84	120.12	171.64	657.70	183.18	61.40	161.78	1862.94	182.32	583.82	164.95
1988	283.70	193.64	106.95	185.03	629.45	198.95	59.94	174.46	1873.89	199.40	580.81	179.22
1989	268.91	208.44	93.78	198.42	601.20	214.72	58.48	187.14	1884.84	216.48	577.80	193.49
1990	254.12	223.24	80.61	211.81	572.95	230.49	57.02	199.82	1895.79	233.56	574.79	207.76
1991	239.33	238.04	67.44	225.20	544.70	246.26	55.56	212.50	1906.74	250.64	571.78	222.03
1992	224.54	252.84	54.27	238.59	516.45	262.03	54.10	225.18	1917.69	267.72	568.77	236.30
1993	209.75	267.64	41.10	251.98	488.20	277.80	52.64	237.86	1928.64	284.80	565.76	250.57
1994	194.96	282.44	27.93	265.37	459.95	293.57	51.18	250.54	1939.59	301.88	562.75	264.84
1995	180.17	297.24	14.76	278.76	431.70	309.34	49.72	263.22	1950.54	318.96	559.74	279.11

contd.

YEARS	MARKETS									
	OSMANABAD		KALLAM		PARBHANI		NAIGAON		BEED	
	Trend values of arrivals	Trend values of prices	Trend values of arrivals	Trend values of prices	Trend values of arrivals	Trend values of prices	Trend values of arrivals	Trend values of prices	Trend values of arrivals	Trend values of prices
1981	125.22	88.53	264.66	88.28	152.12	74.89	108.69	83.52	133.02	88.92
1982	121.32	104.05	259.48	104.44	143.79	89.09	108.84	95.23	124.34	105.69
1983	117.42	119.57	254.30	120.60	135.46	103.29	108.99	106.94	115.66	122.46
1984	113.52	135.09	249.12	136.76	127.13	117.49	109.14	118.65	106.98	139.23
1985	109.62	150.61	243.94	152.92	118.80	131.69	109.29	130.36	98.30	156.00
1986	105.72	166.13	238.76	169.08	110.47	145.89	109.44	142.07	89.62	172.77
1987	101.82	181.65	233.58	185.24	102.14	160.09	109.59	153.78	80.94	189.54
1988	97.92	197.17	228.40	201.40	93.81	174.29	109.74	165.49	72.26	206.31
1989	94.02	212.69	223.22	217.56	85.48	188.49	109.89	177.20	63.58	223.08
1990	90.12	228.21	218.04	233.72	77.15	202.69	110.04	188.51	54.90	239.85
1991	86.22	243.73	212.86	249.88	68.82	216.89	110.19	200.62	46.22	256.62
1992	82.32	259.25	207.68	266.04	60.49	231.09	110.34	212.30	37.54	273.39
1993	78.42	274.77	202.50	282.20	52.16	245.29	110.49	224.04	28.86	290.16
1994	74.52	290.29	197.32	298.36	43.83	259.49	110.64	235.75	20.18	306.93
1995	70.62	305.81	192.14	314.52	35.50	273.69	110.79	247.46	11.50	323.70

Source: As indicated in table 1.

Table 8
The Linear Trend Equations for Arrivals and Prices of Kharif Jowar in Regulated Markets in Marathwada

Markets	*Arrivals*		*Price*	
	a	*b*	*A*	*b*
Aurangabad	402.02***	-14.79**	75.24^{NS}	14.80***
	(105.40)	(6.30)	(78.63)	(4.70)
Lasur	212.31***	-13.17***	77.91*	13.39***
	(60.79)	(3.63)	(56.907)	(3.40)
Jalna	855.45***	-28.25*	72.79^{NS}	15.77***
	(306.64)	(18.33)	(69.76)	(4.17)
Partur	71.62***	-1.46^{NS}	73.02*	12.68***
	(22.62)	(1.35)	(53.24)	(3.18)
Parbhani	160.45***	-8.33**	60.69^{NS}	14.20***
	(56.42)	(3.37)	(63.13)	(3.77)
Beed	141.70***	-8.68***	72.15^{NS}	16.77***
	(37.59)	(2.25)	(70.13)	(4.19)
Latur	1786.29***	10.95^{NS}	62.76^{NS}	17.08***
	(681.05)	(40.70)	(65.76)	(3.93)
Udgir	604.89**	-3.01^{NS}	65.06^{NS}	14.27***
	(299.35)	(17.88)	(56.50)	(3.38)
Osmanabad	129.12***	-3.90*	73.01^{NS}	15.52***
	(42.32)	(2.53)	(63.28)	(3.78)
Kallam	269.84***	-5.18^{NS}	72.12^{NS}	16.16***
	(88.76)	(5.30)	(70.98)	(4.24)
Naigaon	108.54**	0.15^{NS}	71.81^{NS}	11.71***
	(48.58)	(2.90)	(59.74)	(3.57)

Note: Bracketed figures are standard errors.
a = Intercept, b = Slope,
*** Significant at 1 per cent level, ** Significant at 5 per cent level,
* Significant at 10 per cent level, NS = Not-significant.

jowar have increased at the highest annual rate in Latur market and at the lowest rate in Naigaon market during the study period. The annual rate of increase in arrivals of kharif jowar is found to be statistically insignificant in case of both Latur and Naigaon markets.

Among the markets in nine markets annual trend values of kharif jowar arrivals have shown a decreasing trend during the study period. These markets are: Aurangabad, Lasur, Jalna, Partur, Udgir, Osmanabad, Kallam, Parbhani and Beed. The highest base year value of arrival of kharif jowar (i.e. intercept) is found in case of Jalna (85545 quintals) and the lowest in case of Partur (7162 quintals) market. Further, the annual rate of decrease in arrivals of kharif jowar is found to be the highest in case of Jalna (-2825 quintals) and the lowest in case of Partur (-146 quintals) market during the study period. The decreasing trend found in kharif jowar arrivals in most of the markets is due to shift in the area under cultivation from this crop to other crops. The annual rate of decrease in arrivals of kharif jowar is found to be statistically significant at 1 per cent level in case of Lasur and Beed, at 5 per cent level in case of Parbhani and Aurangabad, at 10 per cent level in case of Osmanabad and Jalna markets and statistically insignificant in case of Partur, Udgir and Kallam markets.

The annual trend values of prices of kharif jowar have shown an increasing trend in all the markets. The highest base year value of price of kharif jowar (i.e. intercept) is found in case of Lasur (Rs. 77.91) and the lowest in case of Parbhani (Rs. 60.69) market. The annual rate of increase in the price of kharif jowar is found to be the highest in case of Latur (Rs. 17.08) market and the lowest for Naigaon (Rs. 11.71) market during the period under consideration. Further, the annual rate of increase in kharif jowar price is statistically significant at 1 per cent level in all the markets.

Tur

The annual trend values of arrivals and prices of tur for regulated markets are given in table 9. It is observed from this table that annual trend values of arrivals of tur have shown a rising trend over the period of this study in nine regulated markets viz., Aurangabad, Lasur, Jalna, Partur, Udgir, Osmanabad, Naigaon, Manwat and Beed. Among the markets where annual trend values of arrivals of tur showed rising trend the highest base year value of arrivals of this commodity (i.e. intercept) is found in case of Jalna (57868 quintals) market and the lowest in case of Aurangabad (-2361 quintals) market. The annual rate of increase is found to be the highest in case of Aurangabad (3861 quintals) market and the lowest in case of Manwat (23 quintals) market. It means that the arrivals of tur have increased at the highest annual rate in Aurangabad market and at the lowest rate in Manwat market during the study period. Due to better production of the crop and attractive prices in the market, has resulted in increased arrivals. The annual rate of increase in arrivals of tur is found to be statistically significant at 1 per cent level in case of Lasur and Aurangabad markets, and at 5 per cent level in case of

Table 9

Trend Values of Arrivals and Prices of Tur at Regulated Markets in Marathwada

(Arrivals in '00 Quintals)
(Price in Rs. per Quintal

YEARS	MARKETS													
	AURANGABAD		LASUR		JALNA		PARTUR		LATUR		UDGIR		OSMANABAD	
	Trend values of arrivals	Trend values of prices	Trend values of arrivals	Trend values of prices	Trend values of arrivals	Trend values of prices	Trend values of arrivals	Trend values of prices	Trend values of arrivals	Trend values of prices	Trend values of arrivals	Trend values of prices	Trend values of arrivals	Trend values of prices
1981	15.00	176.68	42.59	134.00	605.70	202.65	134.54	227.67	1595.95	208.72	580.76	211.99	62.46	230.25
1982	53.61	248.83	53.20	211.03	632.72	277.40	140.56	292.13	1586.37	286.25	589.22	278.96	66.36	299.20
1983	92.22	320.98	63.81	288.06	659.74	352.15	146.58	356.59	1576.98	363.78	597.68	345.93	70.26	368.15
1984	130.83	393.13	74.42	365.09	686.76	426.90	152.60	421.05	1567.59	441.31	606.14	412.90	74.16	437.10
1985	169.44	465.28	85.03	442.12	713.78	501.65	158.62	485.51	1558.20	518.84	614.60	479.87	78.06	506.05
1986	208.05	537.43	95.64	519.15	740.80	576.40	164.64	549.97	1548.81	596.37	623.06	546.84	81.96	575.00
1987	246.66	609.58	106.25	596.18	767.82	651.15	170.66	614.43	1539.42	673.90	631.52	613.81	85.86	643.95
1988	285.27	681.73	116.86	673.21	794.84	725.90	176.68	678.89	1530.03	751.43	639.98	680.78	89.76	712.90
1989	323.88	753.88	127.47	750.24	821.86	800.65	182.70	743.35	1520.64	828.96	648.44	747.75	93.66	781.85
1990	362.49	826.03	138.08	827.27	848.88	875.40	188.72	807.81	1511.25	906.49	656.90	814.72	97.56	850.80
1991	401.10	898.18	148.69	904.30	875.90	950.15	194.74	872.27	1501.86	984.02	665.36	881.69	101.46	919.75
1992	439.71	970.33	159.30	981.33	902.92	1024.90	200.76	936.73	1492.47	1061.55	673.82	948.66	105.36	988.70
1993	478.32	1042.48	169.91	1058.36	929.94	1099.65	206.78	1001.19	1483.08	1139.08	682.28	1015.63	109.26	1057.65
1994	516.93	1114.63	180.52	1135.39	956.96	1174.40	212.80	1065.65	1473.69	1216.61	690.74	1082.60	113.16	1126.60
1995	555.54	1564.67	191.13	1212.42	983.98	1249.15	218.82	1130.11	1464.30	1294.14	699.20	1149.57	117.06	1195.55

contd.

YEARS	MARKETS													
	KALLAM		PARBHANI		MANWAT		NANDED		NAIGAON		BEED		MANJLEGAON	
	Trend values of arrivals	*Trend values of prices*	*Trend values of arrivals*	*Trend values of prices*	*Trend values of arrivals*	*Trend values of prices*	*Trend values of arrivals*	*Trend values of prices*	*Trend values of arrivals*	*Trend values of prices*	*Trend values of arrivals*	*Trend values of prices*	*Trend values of arrivals*	*Trend values of prices*
1981	98.68	216.34	295.69	191.11	270.41	177.76	227.50	220.25	18.31	260.49	215.10	189.07	273.30	198.13
1982	96.06	281.99	292.89	262.25	270.64	248.52	223.57	290.33	18.72	295.39	223.41	265.19	272.61	268.81
1983	93.44	347.64	290.09	333.39	270.87	319.28	219.64	360.41	19.13	330.29	231.72	341.31	271.92	339.49
1984	90.82	413.29	287.29	404.53	271.10	390.04	215.71	430.49	19.54	365.19	240.03	417.43	271.23	410.17
1985	88.20	478.94	284.49	475.67	271.33	460.80	211.78	500.57	19.95	400.09	248.34	493.55	270.54	480.85
1986	85.58	544.59	281.69	546.81	271.56	531.56	207.85	570.65	20.36	434.99	256.65	569.67	269.85	551.53
1987	82.96	610.24	278.89	617.95	271.79	602.32	203.92	640.73	20.77	469.89	264.96	645.79	269.16	622.21
1988	80.34	675.89	276.09	689.09	272.02	673.08	199.99	710.81	21.18	504.79	273.27	721.91	268.47	692.89
1989	77.72	741.54	273.29	760.23	272.25	743.84	196.06	780.89	21.59	539.69	281.58	798.03	267.78	763.57
1990	75.10	807.19	270.49	831.37	272.48	814.60	192.13	850.97	22.00	574.59	289.89	874.15	267.09	834.25
1991	72.48	872.84	267.69	902.51	272.71	885.36	188.20	921.05	22.41	609.49	298.20	950.27	266.40	904.93
1992	69.86	938.49	264.89	973.65	272.94	956.12	184.27	991.13	22.82	644.39	306.51	1026.39	265.71	975.61
1993	67.24	1004.14	262.09	1044.79	273.17	1026.88	180.34	1061.21	23.23	679.29	314.82	1102.51	265.02	1046.29
1994	64.62	1069.79	259.29	1115.93	273.40	1097.64	176.41	1131.29	23.64	714.19	323.13	1178.63	264.33	1116.97
1995	62.00	1135.44	256.49	1187.07	273.63	1168.49	172.48	1201.37	24.05	749.09	331.44	1254.75	263.64	1187.65

Source: As indicated in table 1.

Osmanabad market. The annual rate of increase in arrivals of tur is not found statistically significant in case of Jalna, Partur, Udgir, Manwat, Naigaon and Beed markets.

Annual trend values of arrivals of tur have shown a decreasing trend in Kallam, Parbhani, Manjlegaon, Nanded and Latur markets during the period of this study. The highest base year value of arrivals of tur (i.e. intercept) is found in case of Latur (160515 quintals) market and the lowest in case of Kallam (10130 quintals) market. The annual rate of decrease in arrivals of tur is found to be the highest in case of Latur (-933 quintals) market and the lowest in case of Manjlegaon (-69 quintals) during the study period. The annual rate of decrease in arrivals of tur is found statistically insignificant in case of Kallam, Latur, Manjlegaon, Parbhani and Nanded markets.

The annual trend values of prices of tur have shown an increasing trend in all the markets during the study period. The highest base year value of price of this commodity (i.e. intercept) is found in case of Naigaon (Rs. 255.59) and the lowest in case of Lasur (Rs. 56.97) market. The annual rate of increase in the price of tur is found to be the highest in case of Latur (Rs. 77.53) market and the lowest for Naigaon (Rs. 34.70) during the study period. Further, the annual rate of increase in the prices is statistically significant at 1 per cent level in all the markets (see table 10).

Greengram

The annual trend values of arrivals and prices of greengram for regulated markets are given in table 11. It is observed from this table that annual trend values of arrivals of greengram have shown a rising trend over the period of this study in three regulated markets viz., Partur, Udgir and Manwat. Among the markets where annual trend values of arrivals showed rising trend the highest base year value of arrivals (i.e. intercept) is found in case of Manwat (14581 quintals) market and the lowest in case of Udgir (7768 quintals) market. The annual rate of increase in arrivals of this commodity is found to be the highest in case of Partur (405 quintals) and the lowest in case of Manwat (297 quintals) market during the study period. It means that the arrivals of greengram have increased at the highest annual rate in Partur market and at the lowest rate in Manwat market during the study period. The annual rate of increase in arrivals of greengram is

Table 10
The Linear Trend Equations for Arrivals and Prices of Tur in Regulated Markets in Marathwada

Markets	*Arrivals*		*Price*	
	a	*b*	*A*	*b*
Aurangabad	-23.61^{NS}	38.61***	104.53^{NS}	72.15***
	(101.41)	(6.06)	(142.47)	(8.51)
Lasur	31.98^{NS}	10.61***	56.97^{NS}	77.03***
	(67.01)	(4.00)	(150.97)	(9.02)
Jalna	578.68*	27.02^{NS}	127.90^{NS}	74.75***
	(336.47)	(20.11)	(141.39)	(8.45)
Partur	128.52*	6.02^{NS}	163.21*	64.46***
	(94.13)	(5.63)	(116.04)	(6.93)
Parbhani	298.49***	-2.80^{NS}	119.97^{NS}	71.14***
	(100.99)	(6.04)	(118.41)	(7.06)
Manwat	270.18**	0.23^{NS}	107.00^{NS}	70.76***
	(120.68)	(7.21)	(138.55)	(8.28)
Beed	206.79*	8.31^{NS}	112.95^{NS}	76.12***
	(139.52)	(8.34)	(156.57)	(9.36)
Manjlegaon	273.99**	-0.69^{NS}	127.45^{NS}	70.68***
	(110.70)	(6.62)	(118.87)	(7.10)
Osmanabad	58.56**	3.90**	161.30^{NS}	68.95***
	(30.03)	(1.79)	(126.91)	(7.58)
Kallam	101.30**	-2.62^{NS}	150.69^{NS}	65.65***
	(41.78)	(2.50)	(117.35)	(7.01)
Latur	1605.15***	-9.33^{NS}	131.19^{NS}	77.53***
	(553.01)	(33.05)	(157.97)	(9.44)
Udgir	572.30**	8.46^{NS}	126.79^{NS}	77.22***
	(311.99)	(18.64)	(159.08)	(9.51)
Nanded	231.43**	-3.93^{NS}	150.17^{NS}	70.08***
	(94.59)	(5.65)	(133.86)	(8.00)
Naigaon	17.90*	0.41^{NS}	225.59***	34.90***
	(11.10)	(0.66)	(66.05)	(3.95)

Note: Bracketed figures are standard errors.
a = Intercept, b = Slope,
*** Significant at 1 per cent level, ** Significant at 5 per cent level,
* Significant at 10 per cent level, NS = Not-significant.

found to be statistically significant in case of Partur market only at 10 per cent level. The annual rate of increase in arrivals of greengram is not found statistically significant in case of Manwat and Udgir markets. Annual trend values of arrivals of greengram have shown a decreasing trend in Aurangabad, Kallam, Jalna, Latur, Parbhani and Nanded markets during the study period. Among these markets the highest base year value of arrivals of greengram (i.e. intercept) is

Table 11

Trend Values of Arrivals and Prices of Greengram at Regulated Markets in Marathwada

(Arrivals in '00 Quintals)
(Price in Rs. per Quintal)

YEARS	MARKETS									
	AURANGABAD		JALNA		PARTUR		LATUR		UDGIR	
	Trend values of arrivals	Trend values of prices	Trend values of arrivals	Trend values of prices	Trend values of arrivals	Trend values of prices	Trend values of arrivals	Trend values of prices	Trend values of arrivals	Trend values of prices
1981	93.60	193.17	1340.88	206.07	85.68	217.19	449.45	184.66	81.70	150.13
1982	92.71	242.58	1309.63	265.03	89.73	259.10	434.84	246.60	85.72	214.00
1983	91.82	291.99	1278.38	323.99	93.78	301.01	420.23	308.54	89.74	277.87
1984	90.93	341.40	1247.13	382.95	97.83	342.92	405.62	370.48	93.76	341.74
1985	90.04	390.81	1215.88	441.91	101.88	384.83	391.01	432.42	97.78	405.61
1986	89.15	440.22	1184.63	500.87	105.93	426.74	376.40	494.36	101.80	469.48
1987	88.26	489.63	1153.38	559.83	109.98	468.65	361.79	556.30	105.82	533.35
1988	87.37	539.04	1122.13	618.79	114.03	510.56	347.18	618.24	109.84	597.22
1989	86.48	588.45	1090.88	677.75	118.08	552.47	332.57	680.18	113.86	661.09
1990	85.59	637.86	1059.63	736.71	122.13	594.38	317.96	742.12	117.88	724.96
1991	84.70	687.27	1028.38	795.67	126.18	636.29	303.35	804.06	121.90	788.83
1992	83.81	736.68	997.13	854.63	130.23	678.20	288.74	866.00	125.92	852.70
1993	82.92	786.09	965.88	913.59	134.28	720.11	274.13	927.94	129.94	916.57
1994	82.03	835.50	934.63	972.55	138.33	762.02	259.52	989.88	133.96	980.44
1995	81.14	884.91	903.38	1031.51	142.38	803.93	244.91	1051.82	137.98	1044.31

contd.

YEARS	MARKETS							
	KALLAM		*PARBHANI*		*MANWAT*		*NANDED*	
	Trend values of arrivals	*Trend values of prices*	*Trend values of arrivals*	*Trend values of prices*	*Trend values of arrivals*	*Trend values of prices*	*Trend values of arrivals*	*Trend values of prices*
1981	14.00	202.88	531.08	177.54	148.78	183.57	96.98	221.33
1982	13.60	247.27	511.06	235.10	151.75	234.86	92.81	275.94
1983	13.20	291.66	491.04	292.65	154.72	286.15	88.64	330.55
1984	12.80	336.05	471.02	350.22	157.69	337.44	84.47	385.16
1985	12.40	380.44	451.00	407.78	160.66	388.73	80.30	439.77
1986	12.00	424.83	430.98	465.34	163.63	440.02	76.13	494.38
1987	11.60	469.22	410.96	522.90	166.60	491.31	71.96	548.99
1988	11.20	513.61	390.94	580.46	169.57	542.60	67.79	603.60
1989	10.80	558.00	370.92	638.02	172.54	593.89	63.62	658.21
1990	10.40	602.39	350.90	695.58	175.51	645.18	59.45	712.82
1991	10.00	646.78	330.88	753.14	178.48	696.47	55.28	767.43
1992	9.60	691.17	310.86	810.70	181.45	747.76	51.11	822.04
1993	9.20	735.56	290.84	868.26	184.42	799.05	46.94	876.65
1994	8.80	779.95	270.82	925.82	187.39	850.34	42.77	931.26
1995	8.40	824.34	250.80	983.38	190.36	901.63	38.60	985.87

Source: As indicated in table 1.

Table 12
The Linear Trend Equations for Arrivals and Prices of Greengram in Regulated Markets In Marathwada

Markets	*Arrivals*		*Prices*	
	a	*b*	*a*	*b*
Aurangabad	94.49*	-0.89^{NS}	143.76^{NS}	49.41***
	(63.68)	(3.81)	(134.13)	(8.02)
Jalna	372.13***	-31.25^{NS}	147.11^{NS}	58.96***
	(426.82)	(25.51)	(114.23)	(6.83)
Partur	81.63**	4.05*	175.28**	41.91***
	(39.87)	(2.38)	(82.82)	(4.95)
Parbhani	551.10**	-20.02^{NS}	119.98^{NS}	57.56***
	(251.36)	(15.02)	(108.91)	(6.31)
Manwat	145.81^{NS}	2.97^{NS}	132.28*	51.29***
	(122.07)	(7.29)	(84.41)	(5.04)
Latur	464.06**	-14.61*	122.72^{NS}	61.94***
	(177.52)	(10.61)	(109.55)	(6.55)
Udgir	77.68^{NS}	4.02^{NS}	86.26^{NS}	63.87***
	(62.49)	(3.73)	(121.90)	(7.28)
Kallam	14.40**	-0.40^{NS}	158.49**	44.39***
	(6.50)	(0.39)	(66.72)	(3.99)
Nanded	101.55***	-4.17**	166.72**	54.61***
	(37.22)	(2.22)	(93.93)	(5.61)

Note: Bracketed figures are standard errors.
a = Intercept, b = Slope,
*** Significant at 1 per cent level, ** Significant at 5 per cent level,
* Significant at 10 per cent level, NS = Not-significant.

found in case of Jalna (i.e. 137213 quintals) and the lowest for Kallam (14409 quintals) market. The annual rate of decrease in arrivals of greengram is found to be the highest in case of Jalna (-3125 quintals) market and the lowest in case of Kallam (-40 quintals) during the study period. Late arrival of monsoons adversely affects the area under cultivation of this crop which ultimately reduces the production and market arrivals. The annual rate of decrease in arrival of greengram is found to be statistically insignificant in case of Aurangabad, Jalna, Parbhani and Kallam markets. But it is found statistically significant in Latur (at 10 per cent level) and Nanded (at 5 per cent level) markets.

The annual trend values of prices of greengram have shown an increasing trend in all the markets during the study period. The highest base year price (i.e. intercept) is found in case of Partur (Rs. 175.28 per quintal) market and the lowest in case of Udgir (Rs. 86.26

per quintal) market. The annual rate of increase in the price of greengram is found to be the highest in case of Udgir (Rs. 63.87 per quintal) market and the lowest for Partur (Rs. 41.91 per quintal) market during the study period. Further, the annual rate of increase in prices of greengram is statistically significant at 1 per cent level in all the markets.

SECTION B : TRENDS IN ARRIVALS AND PRICES OF OILSEEDS

This section deals with the trend of annual arrivals and prices of oilseeds, namely , safflower, sunflower and groundnut.

Safflower

The annual trend value of arrivals and prices of safflower are shown in table 13. From this table we can see that the annual trend values of arrivals of safflower have shown a decreasing trend in all the eleven markets considered in case of this commodity. It means that the annual arrivals of safflower have decreased in all the markets over the period under consideration. This particular result may be due to low production of this commodity in the areas of the regulated markets in Marathwada during the study period. The base year value of the arrival of safflower is found to be the highest for Jalna (i.e. 60461 quintals) market and the lowest for Osmanabad (i.e. 946 quintals) market. Though the trend values of arrivals of safflower have decreased in all the markets, the annual rate of decrease in the arrivals of safflower is found in case of Jalna (i.e. -3107 quintals) market and the lowest for Osmanabad (i.e. -47 quintals) market. The highest rate of decrease in arrivals of safflower at Jalna market may be due to fast decrease in production of this commodity in the surrounding area of this market. Due to insufficient monsoons, the acreage under cultivation gets reduced and it finally reduces the production which adversely affects the market arrivals. The rate of annual decrease in arrivals of safflower is found statistically significant in all the markets except Lasur and Latur markets. It is significant at 1 per cent level in case of Aurangabad, Jalna, Partur, Parbhani, Manwat, Beed, Manjlegaon and Kallam markets and at 5 per cent level in case of Osmanabad market.

Table 13
Trend Values of Arrivals and Prices of Safflower at Regulated Markets in Marathwada

(Arrivals in '00 Quintals)
(Price in Rs. per Quintal)

YEARS	MARKETS											
	AURANGABAD		LASUR		JALNA		PARTUR		LATUR		OSMANABAD	
	Trend values of arrivals	Trend values of prices	Trend values of arrivals	Trend values of prices	Trend values of arrivals	Trend values of prices	Trend values of arrivals	Trend values of prices	Trend values of arrivals	Trend values of prices	Trend values of arrivals	Trend values of prices
1981	162.70	302.71	139.41	294.39	573.54	305.67	82.98	338.16	109.89	283.15	8.99	324.21
1982	153.79	354.04	134.58	348.11	542.47	357.64	78.26	374.25	108.06	333.97	8.52	257.36
1983	144.88	405.37	129.75	401.83	511.40	409.61	73.54	410.34	106.23	384.79	8.05	390.51
1984	135.97	456.70	124.92	455.55	480.33	461.58	68.82	446.43	104.40	435.61	7.58	423.66
1985	127.06	508.03	120.09	509.27	449.26	513.55	64.10	482.52	102.57	486.43	7.11	456.81
1986	118.15	559.36	115 26	562.99	418.19	565.52	59.38	518.61	100.74	537.25	6.64	489.96
1987	109.24	610.69	110.43	616.71	387.12	617.49	54.66	554.70	98.91	588.07	6.17	523.11
1988	100.33	662.02	105.60	670.4 3	356.05	669.46	49.94	590.79	97.08	638.89	5.70	556.26
1989	91.42	713.35	100.77	724.15	324.98	721.43	45.22	626.88	95.25	689.71	5.23	589.41
1990	82.51	764.68	95.94	777.87	293.91	773.40	40.50	662.97	93.42	740.53	4.76	622.56
1991	73.60	816.01	91.11	831.59	262.84	825.37	35.78	699.06	91.59	791.35	4.29	655.71
1992	64.69	867.34	86.28	885.31	231.77	877.34	31.06	735.15	89.76	842.17	3.82	688.46
1993	55.78	918.67	81.45	939.03	200.70	929.31	26.34	771.24	87.93	892.99	3.35	722.01
1994	46.37	970.00	76.62	992.75	169.63	981.28	21.62	807.33	86.10	943.81	2.88	755.16
1995	37.96	1021.33	71.79	1046.47	138.56	1033.25	16.90	843.42	84.27	994.63	2.41	788.81

contd.

YEARS	MARKETS									
	KALLAM		PARBHANI		MANWAT		BEED		MANJLEGAON	
	Trend values of arrivals	*Trend values of prices*	*Trend values of arrivals*	*Trend values of prices*	*Trend values of arrivals*	*Trend values of prices*	*Trend values of arrivals*	*Trend values of prices*	*Trend values of arrivals*	*Trend values of prices*
1981	51.84	278.10	144.14	366.50	321.53	351.37	197.40	332.81	204.74	303.76
1982	47.90	326.26	134.21	397.67	298.07	383.83	184.26	372.20	192.04	352.44
1983	43.96	374.42	124.28	428.84	274.61	416.29	171.12	411.59	179.34	401.12
1984	40.02	422.58	114.35	460.01	251.15	448.75	157.98	450.98	166.64	449.80
1985	36.08	470.74	104.42	491.18	227.69	481.21	144.84	490.37	153.94	498.48
1986	32.14	518.90	94.49	522.35	204.23	513.67	131.70	529.76	141.24	547.16
1987	28.20	567.06	84.56	553.52	180.77	546.13	118.56	569.15	128.54	595.84
1988	24.26	615.22	74.63	584.69	157.31	578.59	105.42	608.54	115.84	644.52
1989	20.32	663.38	64.70	615.86	133.85	611.05	92.28	647.93	103.14	693.20
1990	16.38	711.54	54.77	647.03	110.39	643.51	79.14	687.32	90.44	741.88
1991	12.44	759.70	44.84	678.20	86.93	675.97	66.00	726.71	77.74	790.56
1992	8.50	807.86	34.91	709.37	63.47	708.43	52.86	766.10	65.04	839.24
1993	4.56	856.02	24.98	740.54	40.01	740.89	39.72	805.49	52.34	887.92
1994	0.62	904.18	15.05	771.71	16.55	773.35	26.58	844.88	39.64	936.60
1995	-3.32	952.34	5.12	802.88	-6.91	805.81	13.44	884.27	26.94	985.28

Source: As indicated in table 1.

It can be seen from table 13 that the annual trend values of prices of safflower have shown a rising trend in all the eleven regulated markets considered in case of this commodity. It means that the price of safflower has increased in all the regulated markets in Marathwada during the period of this study. The base year price of safflower is as low as Rs. 229.94 per quintal in Kallam market and as high as Rs. 335.33 per quintal in Parbhani market. The rate of increase in prices of safflower differed from market to market. However, it is found to be the highest in case of Latur (Rs. 53.72 per quintal) market and the lowest (Rs. 31.17 per quintal) in case of Parbhani market. Further, it is important to note that this rate of increase in price of safflower is found highly significant (at 1 per cent level) in all the regulated markets under consideration.

Table 14
The Linear Trend Equations for Arrivals and Prices of Safflower in Regulated Markets In Marathwada

Markets	*Arrivals*		*Prices*	
	a	*B*	*a*	*b*
Aurangabad	171.61***	-8.91***	251.33**	51.33***
	(43.77)	(2.62)	(102.12)	(6.10)
Lasur	144.24**	-4.83NS	240.61**	53.72***
	(69.10)	(4.13)	(103.90)	(6.21)
Jalna	604.61***	-31.07***	253.70**	51.97***
	(175.88)	(10.51)	(99.08)	(5.92)
Partur	87.70***	-4.72***	302.07***	36.09***
	(28.14)	(1.68)	(107.62)	(6.43)
Parbhani	154.07***	-9.93***	335.33***	31.17***
	(40.17)	(2.40)	(124.50)	(7.44)
Manwat	344.99***	-23.46***	318.91***	32.46***
	(49.62)	(2.97)	(100.78)	(6.02)
Beed	210.54***	-13.14***	293.42***	39.39***
	(47.23)	(2.82)	(77.76)	(4.65)
Manjlegaon	217.44***	-12.70***	255.08***	48.88***
	(38.33)	(2.29)	(94.66)	(5.66)
Latur	111.72***	-1.83NS	232.33**	50.82***
	(31.88)	(2.22)	(99.22)	(5.93)
Osmanabad	9.46**	-0.47**	291.06***	33.15***
	(4.38)	(0.26)	(99.67)	(5.96)
Kallam	55.78***	-3.94***	229.94***	48.16***
	(7.68)	(0.46)	(68.30)	(4.08)

Note: Bracketed figures are standard errors.
a = Intercept, b = Slope,
*** Significant at 1 per cent level, ** Significant at 5 per cent level,
* Significant at 10 per cent level, NS = Not-significant.

Sunflower

The annual trend values of arrivals and prices of sunflower at six regulated markets considered are shown in table 15. It could be seen from the table that the arrivals of sunflower have shown an increasing trend in all the six regulated markets. It means the arrivals of this commodity have shown a rising tendency in all the markets under consideration during the period of this study. The base year value of arrival of this commodity is found to be the highest in case of Kallam (7047 quintals) and lowest for Latur (-5879 quintals) market. The annual rate of increase in arrivals of sunflower is found to be the highest in case of Latur (11219 quintals) and the lowest in case of Osmanabad (350 quintals) market during the period under study. It means the arrivals of sunflower have increased at a faster rate in Latur market and at a slower rate in Osmanabad market. Due to the better prices and better yield per hectare, farmers are stressing on this crop particularly from 1980 onwards. This has ultimately resulted in constant increase in production and increased market arrivals. It is to be noted that the annual rate of increase in arrivals of this commodity is found statistically significant in case of all the six markets considered. It is significant at 1 per cent level in case of Latur, Beed, Manjlegaon and Parbhani markets, at 5 per cent level in case of Osmanabad and Kallam markets.

From table 15 it can be seen that the prices of sunflower have also shown a rising trend in all the six markets considered. The base year price of this commodity is found to be the highest in case of Osmanabad (Rs. 300.32 per quintal) and the lowest in case of Parbhani (Rs. 215.04 per quintal) market. The annual rate of increase in price of this commodity is found to be the highest in case of Beed (Rs. 64.00 per quintal) and the lowest in case of Osmanabad (Rs. 57.43 per quintal) market during the period under study. Further, it is important to note that the annual rate of increase in price of sunflower is statistically significant in all the six markets. It is significant at 1 per cent level in Latur, Osmanabad, Parbhani, Beed, Kallam and Manjlegaon markets.

Groundnut

The annual trend values of arrivals and prices of groundnut are shown in table 17. From this table it is clear that the arrivals of groundnut

Table 15

Trend Values of Arrivals and Prices of Sunflower at Regulated Markets in Marathwada

(Arrivals in '00 Quintals)
(Price in Rs. per Quintal)

YEARS	LATUR		OSMANABAD		KALLAM		MARKETS PARBHANI		BEED		MANJLEGAON	
	Trend values of arrivals	Trend values of prices	Trend values of arrivals	Trend values of prices	Trend values of arrivals	Trend values of prices	Trend values of arrivals	Trend values of prices	Trend values of arrivals	Trend values of prices	Trend values of arrivals	Trend values of price
1981	53.40	341.35	35.45	357.75	84.30	313.30	-4.54	273.83	86.92	293.44	-0.62	315.48
1982	165.59	403.84	38.95	415.18	98.13	375.68	5.25	332.62	104.28	357.44	28.16	375.35
1983	277.78	466.33	42.45	472.61	101.96	438.06	15.04	391.41	121.64	421.44	56.94	435.22
1984	389.97	528.82	45.95	530.04	105.79	500.44	24.83	450.20	139.00	485.44	85.72	495.09
1985	502.16	591.31	49.45	587.47	109.62	562.82	34.62	508.99	156.36	549.44	114.50	554.96
1986	614.35	653.80	52.95	644.90	113.45	625.20	44.41	567.78	173.72	613.44	143.28	614.83
1987	726.54	716.29	56.45	702.33	117.28	687.58	54.20	626.57	191.08	6744.44	172.06	674.70
1988	838.73	778.78	59.95	759.76	121.11	749.96	63.99	685.36	208.44	741.44	200.84	734.57
1989	950.92	841.27	63.45	817.19	124.94	812.34	73.78	744.15	225.80	805.44	229.62	794.44
1990	1063.11	903.76	66.95	874.62	128.77	874.72	83.57	802.94	243.16	869.44	258.40	854.31
1991	1175.30	966.25	70.45	932.05	132.60	937.10	93.36	861.73	260.52	933.44	287.18	914.18
1992	1287.49	1028.74	73.95	989.48	136.43	999.48	103.15	920.52	277.88	997.44	315.96	974.05
1993	1399.68	1091.23	77.45	1046.91	140.26	1061.86	112.94	979.31	295.24	1061.44	344.74	1033.92
1994	1511.87	1153.72	80.95	1104.34	144.09	1124.24	122.73	1038.10	312.60	1125.44	373.52	1093.79
1995	1624.06	1216.21	84.45	1161.77	147.92	1186.62	132.52	1096.89	329.96	1189.44	402.30	1153.66

Source: As indicated in table 1.

Table 16
The Linear Trend Equations for Arrivals and Prices of Sunflower in Regulated Markets In Marathwada

Markets	*Arrivals*		*Price*	
	a	*b*	*A*	*b*
Parbhani	-14.33^{NS} (59.59)	9.79*** (3.56)	215.04** (103.89)	58.79*** (6.21)
Beed	69.56^{NS} (108.59)	17.36*** (6.49)	229.44** (112.74)	64.00*** (6.74)
Manjlegaon	-29.40^{NS} (62.37)	28.78*** (3.73)	255.61*** (92.00)	59.87*** (5.50)
Latur	-58.79^{NS} (542.00)	112.19*** (32.39)	278.86*** (99.15)	62.49*** (5.93)
Osmanabad	31.95^{NS} (25.15)	3.50** (1.50)	300.32*** (91.66)	57.43*** (5.48)
Kallam	70.47** (33.47)	3.83** (2.00)	250.92** (104.40)	62.38*** (6.24)

Note: Bracketed figures are standard errors.
a = Intercept, b = Slope,
*** Significant at 1 per cent level, ** Significant at 5 per cent level,
* Significant at 10 per cent level, NS = Not-significant.

have shown a rising trend in seven markets viz.; Udgir, Kallam, Parbhani, Manwat, Nanded, Beed and Manjlegaon markets during the period under study. The arrivals of this commodity have shown a decreasing trend in three markets viz.; Jalna, Latur and Osmanabad. Among the markets where the arrivals have shown rising trend, the highest annual rate of increase in arrivals of this commodity is found in case of Manjlegaon (3241 quintals) and the lowest in case of Udgir (11 quintals) market during the period under study. The assured irrigation during summer season has attracted farmers to take this crop on larger scales. Secondly, the prices of groundnut are also attractive. The yield of the crop is further satisfactory which ultimately results in enhanced production and market arrivals. The annual rate of increase in arrivals of this commodity is found statistically significant at 1 per cent level in case of Manjlegaon and at 5 per cent level in case of Parbhani, Manwat and Nanded markets. However, the rates of increase in arrivals for Udgir, Kallam and Beed markets are found to be insignificant.

Table 17
Trend Values of Arrivals and Prices of Groundnut at Regulated Markets in Marathwada

(Arrivals in '00 Quintals)
(Price in Rs. per Quintal)

YEARS	MARKETS									
	JALNA		LATUR		UDGIR		OSMANABAD		KALLAM	
	Trend values of arrivals	Trend values of prices	Trend values of arrivals	Trend values of prices	Trend values of arrivals	Trend values of prices	Trend values of arrivals	Trend values of prices	Trend values of arrivals	Trend values of prices
1981	86.24	206.07	611.15	318.62	7.70	278.82	33.72	321.13	29.27	279.58
1982	80.96	265.03	593.27	383.32	7.81	331.90	32.95	378.63	29.45	342.55
1983	75 68	323.99	575.39	448.02	7.92	384.98	32.18	436.13	29.63	405.52
1984	70.40	382.95	557.51	512.72	8.03	438.06	31.41	493.63	29.81	468.49
1985	65.12	441.91	539.63	577.42	8.14	491.14	30.64	551.13	29.99	531.46
1986	59.84	500.87	521.75	642.12	8.25	544.22	29.87	608.63	30.17	594.43
1987	54.56	559.83	503.87	706.82	8.36	597.30	29.10	666.13	30.35	657.40
1988	49.28	618.79	485.99	771.52	8.47	650.38	28.33	723.63	30.53	720.37
1989	44.00	677.75	468.11	836.22	8.58	703.46	27.56	781.13	30.71	783.34
1990	38.72	736.71	450.23	900.92	8.69	756.54	26.79	838.63	30.89	846.31
1991	33.44	795.67	432.35	965.62	8.80	809.62	26.02	896.13	31.07	909.28
1992	28.16	854.63	414.47	1030.32	8.91	862.70	25.25	953.63	31.25	972.25
1993	22.88	913.59	396.59	1095.02	9.02	91 5.78	24.48	1011.13	31.43	1035.22
1994	17.60	972.55	378.71	1159.72	9.13	968.86	23.71	1068.63	31.61	1098.19
1995	12.32	1031.51	360.83	1224.42	9.24	1021.94	22.94	1126.13	31.79	1161.16

contd.

YEARS	MARKETS									
	PARBHANI		MANWAT		NANDED		BEED		MANJLEGAON	
	Trend values of arrivals	Trend values of prices	Trend values of arrivals	Trend values of prices	Trend values of arrivals	Trend values of prices	Trend values of arrivals	Trend values of prices	Trend values of arrivals	Trend values of prices
1981	6.00	283.20	3.83	295.48	115.73	251.48	164.34	285.37	9.28	297.71
1982	15.45	333.37	14.59	346.30	139.47	319.65	167.10	351.58	41.69	357.27
1983	24.90	383.54	25.35	397.12	163.21	387.82	169.86	417.79	74.10	416.83
1984	34.35	433.71	36.11	447.94	186.95	455.99	172.62	484.00	106.51	476.39
1985	43.80	483.88	46.87	498.76	210.69	524.16	175.38	550.21	138.92	535.95
1986	53.25	534.05	57.63	549.58	234.43	592.33	178.14	616.42	171.33	595.51
1987	62.70	584.22	68.39	600.40	258.17	660.50	180.90	682.63	203.74	655.07
1988	72.15	634.39	79.15	651.22	281.91	728.67	183.66	748.84	236.15	714.63
1989	81.60	684.56	89.91	702.04	305.65	796.84	186.42	815.05	268.56	774.19
1990	91.05	734.73	100.67	752.86	329.39	865.01	189.18	881.26	300.97	833.75
1991	100.50	784.90	111.43	803.68	353.13	933.18	191.94	947.47	333.38	893.31
1992	109.95	835.07	122.19	854.50	376.87	1001.35	194.70	1013.68	365.79	952.87
1993	119.40	885.24	132.95	905.32	400.61	1069.52	197.46	1079.89	398.20	1012.43
1994	128.85	935.41	143.71	956.14	424.35	1137.69	200.22	1146.10	430.61	1071.99
1995	138.30	985.58	154.47	1006.96	448.09	1205.86	202.98	1212.31	463.02	1131.55

Source: As indicated in table 1.

The arrivals of groundnut have shown the highest rate of annual decrease for Latur (-1788 quintals) and the lowest for Osmanabad (-77 quintals). The annual rate of decrease in arrivals for Jalna and Latur markets are statistically significant at 5 per cent level but for Osmanabad it is statistically insignificant.

Table 17 shows that the prices of groundnut have shown a rising trend during the study period in all the ten markets. However, the rate of increase in the price of groundnut differed from market to market. It is found to be the highest in case of Nanded market (Rs. 68.17 per quintal) and the lowest for Parbhani (Rs. 50.17 per quintal) market. It is important to note here that the rate of increase in prices of groundnut is found to be highly significant (at 1 per cent level) in all the markets.

Prices for all the crops in all the markets have showed an increasing trend. This may be attributed to the factors such as population growth, changes in the weather conditions, increase in per capita income of the consumers and rising cost of inputs in farming.

The discussion made in this chapter indicates that the arrivals of wheat commodity have not shown the similar trend in all the markets. The arrivals of wheat have shown a rising trend in seven markets and a falling trend in another seven markets. The annual rate of increase or decrease in arrivals is also found to be different for different markets. The highest and the lowest annual rates of increase in arrivals of wheat are found for Aurangabad and Osmanabad markets respectively whereas the highest and lowest annual rates of decrease in arrivals are found for Latur and Beed markets respectively. Trend in prices of wheat has been increasing in all the markets. The prices of wheat have exhibited the highest annual rate of increase in case of Latur and lowest for Naigaon market. In case of bajra its arrivals have shown a rising trend in seven markets. However, the arrivals of this commodity have shown a falling trend in three markets. Arrivals of bajra have shown the highest annual increase in Jalna market. Annual trend values of arrivals of this commodity have remained stagnant in Lasur market over the period under study. As the bajra crop requires less rainfall and unfortunately the rains in the recent years are poor, farmers have started growing more bajra, so, the bajra arrivals in these markets have showed an increase. The prices of bajra have shown a rising trend in all the markets. Annual increase in the prices of bajra is found to be the highest in Lasur and the lowest in Udgir market.

Rabi jowar arrivals have shown a decreasing trend in nine markets and increasing trend in only four markets. The highest annual rate of decrease in arrivals of rabi jowar is found for Manjlegaon and the lowest for Kallam market. The decreasing trend is observed due to the reduced production of the crop and poor weather conditions. Further, the highest rate of increase in arrivals of rabi jowar is found for Jalna and lowest for Osmanabad market. The prices of this commodity have shown a rising trend in all the markets. The highest annual rate of increase in price of this commodity is found in Lasur and lowest in Nanded market.

In this study it is found that the arrivals of Kharif jowar have shown a decreasing trend in all the markets except Latur and Naigaon markets. The annual rate of decrease is found to be the highest for Jalna and lowest for Partur market. The decreasing trend found in kharif jowar arrivals in most of the markets is due to shift in the area under cultivation from this crop to other crops. Though the prices of Kharif jowar have shown an increasing trend in all the markets, its annual rate of increase differed from market to market (highest for Latur and lowest for Naigaon). The arrivals of tur have shown rising trend in nine markets and falling trend in five markets during the period under study. Its arrivals have recorded the highest annual rate of increase in Aurangabad and lowest rate in Manwat market. Due to better production of the crop and attractive prices in the market, it has resulted in increased arrivals. The annual rate of increase in its prices is found to be the highest for Latur and lowest for Naigaon market. Greengram arrivals have shown a decreasing trend in six markets and the annual rate of increase found to be the highest for Jalna and lowest for Kallam market. Though the prices of this commodity have shown a rising trend, its annual rate of increase has differed from market to market (highest for Udgir and lowest for Partur).

This study has indicated that the arrivals and prices of safflower have shown opposite trends in all the markets. Its arrivals have shown a decreasing trend but its prices have shown an increasing trend during the study period. The highest annual rate of decrease in

Table 18
The Linear Trend Equations for Arrivals and Prices of Groundut in Regulated Markets in Marathwada

Markets	*Arrivals*		*Price*	
	a	*b*	*A*	*b*
Jalna	91.52*** (33.75)	-5.28** (2.02)	147.11^{NS} (114.23)	58.96*** (6.83)
Parbhani	-3.45^{NS} (65.93)	9.45** (3.94)	233.03* (163.95)	50.17*** (9.80)
Manwat	-6.93^{NS} (75.80)	10.76** (4.53)	244.66** (114.45)	50.82*** (6.84)
Beed	161.58** (84.62)	2.76^{NS} (5.06)	219.16** (109.27)	66.21*** (6.53)
Manjlegaon	-23.13^{NS} (88.39)	32.41*** (5.28)	238.18** (102.46)	59.56*** (6.12)
Latur	629.03*** (134.08)	-17.88** (8.01)	253.92** (102.52)	64.70*** (6.13)
Udgir	7.59^{NS} (5.66)	0.11^{NS} (0.34)	225.74* (167.54)	53.08*** (10.01)
Osmanabad	34.49** (14.91)	-0.77^{NS} (0.89)	263.63*** (94.29)	57.50*** (5.64)
Nanded	91.99^{NS} (215.47)	23.74** (12.88)	183.31^{NS} (139.04)	68.17*** (8.31)
Kallam	29.09* (16.51)	0.18^{NS} (0.99)	216.61* (124.24)	62.97*** (7.42)

Note: Bracketed figures are standard errors.
a = Intercept, b = Slope,
*** Significant at 1 per cent level, ** Significant at 5 per cent level,
* Significant at 10 per cent level, NS = Not-significant.

arrivals is found for Jalna and lowest for Osmanabad. But the highest rate of increase in its prices is found for Latur and lowest for Parbhani. Both prices and arrivals of sunflower have shown a positive trend in all the markets during the study period. The annual rate of increase in arrivals is the highest for Latur and lowest for Osmanabad. But in case of prices of sunflower it is the highest for Beed and lowest for Osmanabad. Due to the better prices and better yield per hectare, farmers are stressing on this crop particularly from 1980 onwards. This has ultimately resulted in constant increase in production and increased market arrivals. Groundnut arrivals have shown a rising trend in most of the markets (i.e. seven out of 10 markets). Groundnut arrivals have exhibited the highest annual rate of increase in Manjlegaon and the lowest rate of increase in Udgir market. Prices of groundnut have shown a rising trend in all the markets. They have increased at the highest annual rate in Nanded and at the lowest rate in Parbhani market.

Prices for all the crops in all the markets have showed an increasing trend. This may be attributed to the factors such as population growth, changes in the weather conditions, increase in per capita income of the consumers and rising cost of inputs in farming.

6

RELATIONSHIP BETWEEN ARRIVALS AND PRICES OF AGRICULTURAL COMMODITIES

The study of variation in market arrivals of agricultural products and its relation to price fluctuations over a time is most important for formulating a sound agricultural and marketing policy. There are two opposite views which have been stated by different economists about the relationship between the marketed surplus (arrival) and prices of foodgrains. Mathur and Ezekeil have stated the hypothesis that there is an inverse relationship between marketed surplus (arrival) and prices of foodgrains. Rajkrishna, Dantwala and Schultz have stated the hypothesis that there is a positive relationship between marketed surplus and price. The first hypothesis of negative relationship holds true in the case of small peasant farmers in underdeveloped and developing agricultures. This hypothesis is explained in terms of fixed and low money obligations of these farmers, by Mathur and Ezekeil. Krishna and Olson argued that an increase in the price of subsistence crop may increase the producer's real income sufficiently so that the income effect on his demand for consumption of this crop outweighs the price effects on production and consumption. The marketed surplus or arrival may, therefore, vary inversely with the market price. As far as hypothesis of positive relationship is concerned, Schultz says that the doctrine that farmers in poor countries either are indifferent or respond perversely to changes in price is patently false and harmful. Rajkrishna gives evidence of positive response of Punjab producers to price incentives.

The seasonal nature of agricultural production, low capacity of the small and marginal farmers to hold stock and to obtain credit to postpone their sales, less storability of agricultural products are the factors which explain the variation in arrivals and prices of agricultural commodities from one season to another season within a year. The tendency is that the market supply of agricultural commodities remains very high in peak season and very low in lean season. The excess market arrival of agricultural commodities pushes the prices down to very low level in peak season and low market supply pushes them to unduly high level during lean season. Naturally the peak and lean seasons differ from commodity to commodity.

In this chapter an attempt is made to examine the relationship between monthly arrivals and average prices of foodgrains and oilseeds at regulated markets in the Marathwada region. An attempt is also made in this chapter to examine the seasonal pattern of market arrivals of foodgrains and oilseeds. The relationship between market arrivals and prices of foodgrains and oilseeds is measured by computing correlation coefficient between these two variables. Seasonal pattern is studied by taking the average percentage quantity marketed and average prices of agricultural commodities during three marketing seasons viz., peak, mid and lean season, at regulated markets in Marathwada. The agricultural commodities considered for the analysis include foodgrains (viz., wheat, kharif jowar, rabi jowar, bajra, greengram and tur) and oilseeds (viz., safflower, sunflower and groundnut).

This chapter is divided into two sections. Section 'A' deals with the relationship between arrivals and prices of foodgrains and oilseeds and section 'B' deals with the seasonal pattern of market arrivals and prices of foodgrains and oilseeds at regulated markets.

SECTION A : RELATIONSHIP BETWEEN ARRIVALS AND PRICES

This section deals with the relationship between arrivals and prices of foodgrains and oilseeds at regulated markets in Marathwada. The commodity-wise analysis is presented below :

Wheat

The coefficients of correlation between monthly arrivals and prices of wheat are presented in table 1. It is clear from this table that prices and arrivals are negatively correlated with each other. The markets where highly negative relationship between monthly prices and arrivals of wheat is found are Aurangabad, Lasur, Manwat, Partur, Jalna, Manjlegaon, Beed, Kallam, Nanded and Latur. In these markets the monthly arrivals and prices of wheat have varied in opposite directions during the period of this study. It means that an increase in arrivals of wheat were accompanied by a fall in its price and vice versa in most of the markets and during most of the years. There are few markets for which the relationship between arrivals and prices of wheat is found to be positive during very few years.

Bajra

The correlation coefficients between monthly arrivals and prices of bajra are presented in table 2. A look at this table gives us a different picture about the relationship between monthly arrivals and prices of bajra during the period under study. Out of ten markets in eight markets (except Beed and Manjlegaon) there has been a positive relationship between arrivals and prices of bajra. It means that in most of the selected markets a highly positive relationship between arrivals and prices is found. These markets are : Partur, Manwat, Udgir, Latur, Aurangabad, Lasur, Jalna and Kallam. This positive relationship indicates that larger supplies of bajra into the market did not cause the price to fall.

Kharif Jowar

In case of Kharif jowar correlation coefficients between monthly arrivals and prices are turned out to be positive in some markets in some years and negative in other markets in other years. Out of eleven markets studied in case of Kharif jowar in eight markets the correlation coefficients are negative and very high. These markets are: Aurangabad, Jalna, Parbhani, Beed, Latur, Udgir, Kallam and Naigaon. Highly positive relationship between arrivals and prices of Kharif jowar is found in case of Latur, Partur and Osmanabad markets. The point to note is that the relationship between arrivals

Table 1
Coefficients of Correlation Between Monthly Prices and Arrivals of Wheat for Different Regulated Markets

	MARKETS						
Years	*Aurangabad*	*Lasur*	*Jalna*	*Partur*	*Parbhani*	*Manwat*	*Beed*
1981	-0.23	-0.22	-0.35	0.14	-0.14	-0.07	-0.51
1982	-0.49	-0.32	-0.67	-0.60	-0.67	-0.79	-0.74
1983	-0.36	-0.49	-0.71	-0.76	0.06	-0.40	-0.72
1984	'-0.32'	-0.19	-0.38	-0.42	-0.36	-0.33	-0.46
1985	-0.73	-0.52	-0.72	-0.63	-0.75	-0.71	-0.73
1986	0.02	-0.46	-0.09	-0.10	0.09	0.74	0.24
1987	-0.59	-0.44	-0.27	-0.49	0.25	-0.45	-0.18
1988	-0.37	-0.38	-0.48	-0.62	-0.02	-0.60	-0.46
1989	-0.55	-0.58	0.37	-0.37	0.43	-0.58	-0.35
1990	-0.68	-0.69	-0.31	-0.58	-0.45	0.04	-0.46
1991	-0.44	0.12	-0.53	-0.58	-0.47	-0.64	-0.65
1992	-0.01	-0.17	0.21	-0.41	0.09	-0.61	-0.53
1993	-0.17	-0.61	-0.68	-0.81	-0.57	-0.69	-0.75
1994	-0.49	-0.18	-0.61	-0.78	0.41	-0.60	-0.78
1995	-0.17	-0.34	0.64	0.48	0.40	0.24	-0.04

	MARKETS						
Years	*Manjlegaon*	*Latur*	*Udgir*	*Osmanabad*	*Kallam*	*Nandad*	*Naigaon*
1981	-0.66	0.24	-0.64	0.15	0.20	-0.51	0.74
1982	-0.83	-0.69	-0.70	-0.52	0.23	-0.59	-0.47
1983	-0.72	-0.76	-0.78	-0.88	0.05	-0.78	-0.70
1984	'-0.32'	-0.07	-0.22	-0.36	-0.54	-0.36	-0.15
1985	-0.59	-0.63	-0.39	-0.47	-0.35	-0.74	-0.51
1986	0.38	0.13	0.14	-0.35	0.36	0.28	0.51
1987	-0.16	0.50	-0.36	0.02	-0.45	-0.25	0.49
1988	-0.35	-0.73	-0.46	-0.60	-0.33	0.18	0.04
1989	0.06	-0.28	0.35	-0.12	-0.09	0.05	-0.26
1990	-0.32	-0.51	-0.30	0.08	-0.54	-0.45	0.15
1991	-0.58	-0.26	-0.04	-0.63	-0.70	0.14	-0.28
1992	-0.66	-0.32	0.38	0.14	-0.27	-0.58	0.27
1993	-0.79	-0.50	0.37	-0.54	-0.73	-0.36	0.11
1994	-0.29	-0.38	-0.25	-0.66	-0.42	-0.38	-0.46
1995	0.19	0.08	0.38	0.12	0.25	-0.36	0.40

Source: Data on monthly prices and arrivals are compiled from the office records of regulated markets.

Table 2
Coefficients of Correlation Between Monthly Prices and Arrivals of Bajra for Different Regulated Markets

			MARKETS		
Years	*Aurangabad*	*Lasur*	*Jalna*	*Partur*	*Manwat*
1981	-0.04	0.96	-0.16	0.11	0.22
1982	0.36	0.91	0.38	0.13	-0.13
1983	-0.57	-0.70	-0.57	-0.39	-0.48
1984	0.56	0.35	0.46	0.58	0.29
1985	-0.37	0.38	0.53	0.51	0.50
1986	-0.46	-0.58	-0.79	0.01	0.16
1987	0.10	-0.44	-0.15	0.27	0.53
1988	0.16	0.19	-0.16	-0.31	0.39
1989	0.54	-0.15	0.43	-0.02	-0.57
1990	0.44	0.57	0.11	0.08	0.42
1991	0.54	-0.04	0.55	0.43	0.55
1992	-0.90	-0.94	-0.89	-0.42	-0.35
1993	0.75	0.40	0.54	0.51	0.06
1994	0.28	0.17	0.40	0.39	0.40
1995	-0.65	-0.25	-0.56	-0.42	0.28

			MARKETS		
Years	*Beed*	*Manjlegaon*	*Latur*	*Udgir*	*Kallam*
1981	-0.57	-0.28	-0.70	-0.60	0.03
1982	0.21	0.23	0.26	0.26	0.01
1983	-0.56	-0.34	-0.33	-0.26	0.57
1984	-0.26	-0.12	0.83	0.82	0.60
1985	0.49	0.24	0.72	0.86	0.56
1986	-0.71	-0.18	-0.21	-0.10	0.37
1987	0.38	0.68	-0.08	0.49	0.08
1988	-0.21	0.17	0.40	-0.06	-0.46
1989	-0.45	-0.25	-0.75	0.09	-0.75
1990	-0.20	0.03	0.60	0.30	0.75
1991	-0.18	-0.12	0.05	0.21	0.76
1992	-0.92	-0.62	-0.80	-0.16	-0.68
1993	0.79	0.40	0.94	0.60	0.89
1994	-0.07	-0.10	0.43	0.49	0.51
1995	-0.82	-0.44	-0.59	0.05	-0.78

Source: As indicated in table 1.

and prices of Kharif jowar has been negative in most of the markets and during most of the years considered in the present study.

Rabi Jowar

The data on correlation coefficients between arrivals and prices of rabi jowar presented in table 4 show that there is a very weak relationship between monthly arrivals and prices of this commodity. These correlation coefficients are turned out to be negative in case of most of the markets. Out of 13 markets in 10 markets a highly negative relationship between arrivals and prices of rabi jowar is found. These markets are : Aurangabad, Jalna, Partur, Parbhani, Beed, Latur, Udgir, Osmanabad, Manwat and Lasur. In the remaining markets, the correlation coefficients have indicated a weak positive relationship. These markets are : Manjlegaon, Kallam and Nanded. In case of these three markets larger supplies of rabi jowar into the market did not cause the price to fall.

Tur

Tur is a principal pulse crop of this area. The table 5 shows that the correlation coefficients between arrivals and prices of tur are high and negative in most of the markets and in most of the years. Out of 14 markets in 9 markets the correlation coefficients have indicated a highly negative relationship in most of the years. These markets are : Jalna, Parbhani, Manwat, Manjlegaon, Latur, Udgir, Osmanabad, Kallam and Nanded. In the remaining markets the correlation coefficients are found to be positive. These markets are Aurangabad, Lasur, Beed, Partur and Naigaon. The above discussion clearly indicates that there has been a negative relationship between arrivals and prices of tur during the period under consideration. There are few markets for which the relationship between arrivals and prices of tur, during a very few years, is found to be positive.

Greengram

In case of greengram the coefficients of correlation between arrivals and prices are turned out to be negative in most of the years and in most of the markets viz., Jalna, Manwat, Latur, Udgir and Nanded markets. But there are some markets where prices and arrivals of greengram are found to be positively correlated. Out of nine markets

Table 3
Coefficients of Correlation Between Monthly Prices and Arrivals of Kharif Jowar for Different Regulated Markets

	Markets					
Years	*Aurangabad*	*Lasur*	*Jalna*	*Partur*	*Parbhani*	*Beed*
1981	-0.77	0.68	-0.91	-0.17	-0.13	-0.55
1982	0.38	0.81	-0.02	-0.04	-0.21	-0.27
1983	-0.52	-0.04	-0.44	0.01	-0.80	-0.14
1984	-0.09	-0.35	-0.86	-0.27	-0.62	-0.61
1985	-0.21	-0.14	0.66	0.06	0.12	-0.30
1986	-0.15	-0.40	-0.79	-0.11	-0.07	-0.39
1987	-0.80	-0.78	-0.84	-0.62	0.14	-0.61
1988	0.58	0.02	0.61	0.71	-0.35	0.39
1989	-0.47	-0.13	-0.80	0.14	0.05	-0.83
1990	-0.41	0.55	-0.14	-0.42	-0.09	0.24
1991	0.18	0.23	-0.26	0.61	0.12	0.83
1992	-0.65	-0.26	-0.93	-0.92	-0.98	-0.64
1993	0.82	0.41	0.74	-0.11	-0.20	-0.12
1994	0.31	0.55	0.07	0.40	0.53	0.78
1995	-0.46	0.57	-0.41	0.10	-0.03	-0.65

	Markets				
Years	*Latur*	*Udgir*	*Osmanabad*	*Kallam*	*Naigaon*
1981	-0.34	-0.52	-0.71	-0.90	-0.21
1982	-0.32	-0.15	-0.42	-0.03	0.07
1983	-0.68	-0.81	0.53	-0.08	-0.75
1984	-0.74	-0.82	-0.11	-0.08	-0.80
1985	0.59	0.77	0.30	0.44	-0.02
1986	-0.63	-0.56	-0.78	-0.53	-0.70
1987	-0.85	-0.67	0.13	-0.83	-0.98
1988	-0.54	-0.67	0.56	-0.23	-0.50
1989	-0.81	-0.72	-0.29	-0.82	-0.73
1990	-0.54	0.70	-0.07	-0.50	-0.30
1991	0.33	-0.14	0.81	0.43	0.49
1992	-0.91	-0.76	-0.88	0.93	-0.87
1993	0.66	0.43	0.34	0.16	0.71
1994	0.48	0.30	0.53	-0.23	0.79
1995	-0.21	-0.78	-0.87	-0.80	0.26

Source: As indicated in table 1.

Table 4
Coefficients of Correlation Between Monthly Prices and Arrivals of Rabi Jowar for Different Regulated Markets

			Markets			
Years	*Aurangabad*	*Lasur*	*Jalna*	*Partur*	*Parbhani*	*Manwat*
1981	0.01	0.27	-0.02	-0.53	-0.05	-0.03
1982	-0.43	0.51	-0.12	-0.66	-0.59	-0.54
1983	-0.72	-0.38	-0.58	-0.67	-0.53	-0.52
1984	-0.58	-0.38	-0.36	0.18	0.19	0.01
1985	-0.49	-0.03	-0.48	-0.49	-0.36	-0.38
1986	-0.16	-0.23	-0.23	0.40	-0.41	0.17
1987	-0.08	0.03	0.14	-0.70	-0.46	-0.57
1988	-0.80	-0.44	-0.53	-0.45	-0.45	-0.63
1989	0.19	0.57	0.39	0.39	0.41	0.82
1990	0.34	-0.72	-0.01	0.53	-0.32	0.25
1991	-0.81	-0.89	-0.76	-0.69	-0.71	-0.52
1992	0.17	0.40	0.46	-0.03	0.32	0.65
1993	0.40	0.20	-0.18	0.37	-0.55	0.16
1994	-0.39	-0.12	-0.17	-0.36	0.22	-0.23
1995	-0.18	0.50	0.44	0.10	0.47	0.63

			Markets				
Years	*Beed*	*Manjlegaon*	*Latur*	*Udgir*	*Osmanabad*	*Kallam*	*Nanded*
1981	-0.17	0.42	-0.03	-0.20	0.16	0.36	-0.34
1982	-0.66	-0.52	-0.11	-0.24	-0.33	-0.22	-0.40
1983	-0.60	-0.49	0.49	-0.28	-0.16	-0.57	0.25
1984	-0.44	0.08	-0.30	0.20	-0.50	0.25	-0.04
1985	-0.63	0.06	-0.63	-0.34	-0.88	-0.03	0.32
1986	-0.09	0.03	-0.47	-0.21	-0.58	0.32	0.09
1987	-0.61	-0.43	-0.74	-0.15	-0.62	0.31	-0.52
1988	-0.45	-0.44	-0.58	0.67	-0.45	-0.46	0.24
1989	0.29	0.68	0.26	0.52	0.35	0.61	-0.38
1990	0.03	0.20	0.08	-0.35	0.03	0.11	0.62
1991	-0.70	-0.41	-0.64	0.03	-0.70	-0.40	-0.12
1992	0.51	0.66	0.08	0.34	0.10	0.59	0.17
1993	-0.37	0.09	-0.59	-0.53	-0.42	-0.03	-0.32
1994	-0.24	-0.28	0.71	0.24	-0.27	0.41	0.41
1995	0.49	0.61	0.09	0.18	0.25	0.60	0.45

Source: As indicated in table 1.

Table 5
Coefficients of Correlation Between Monthly Prices and Arrivals of Tur for Different Regulated Markets

			Markets				
Years	*Aurangabad*	*Lasur*	*Jalna*	*Partur*	*Parbhani*	*Manwat*	*Beed*
1981	0.13	0.29	-0.37	0.15	-0.12	-0.56	0.05
1982	-0.16	-0.003	-0.18	-0.48	-0.34	0.19	0.09
1983	-0.32	0.34	-0.67	-0.42	-0.08	-0.52	0.16
1984	0.12	-0.11	0.05	-0.36	0.25	-0.05	0.06
1985	0.10	0.05	-0.10	-0.29	-0.42	-0.35	0.01
1986	-0.13	0.26	-0.66	-0.62	0.76	-0.53	0.18
1987	0.37	0.30	-0.45	0.15	0.10	0.12	0.17
1988	0.39	0.20	0.40	0.18	0.30	0.19	-0.31
1989	-0.24	-0.63	-0.76	-0.76	-0.89	-0.24	0.18
1990	-0.40	0.04	-0.64	-0.48	-0.63	-0.05	0.65
1991	-0.25	-0.007	-0.29	-0.05	0.51	-0.27	0.16
1992	0.33	0.02	0.05	0.17	0.17	0.27	-0.04
1993	0.47	0.38	0.28	0.14	-0.25	0.19	-0.47
1994	0.37	0.17	-0.27	0.02	-0.52	-0.34	0.14
1995	0.12	-0.18	-0.24	0.21	-0.68	-0.31	0.17

				Markets			
Years	*Manjlegaon*	*Latur*	*Udgir*	*Osmanabad*	*Kallam*	*Nanded*	*Naigaon*
1981	-0.01	-0.52	-0.74	-0.41	-0.05	-0.59	0.41
1982	-0.26	-0.86	-0.60	-0.58	-0.64	-0.69	-0.02
1983	-0.17	-0.77	-0.83	-0.74	-0.58	-0.88	0.25
1984	0.02	0.03	-0.32	0.13	-0.48	-0.10	0.12
1985	-0.34	-0.56	-0.66	-0.29	-0.38	-0.35	-0.63
1986	-0.57	-0.86	-0.88	-0.48	-0.75	-0.87	0.03
1987	0.05	0.07	-0.17	0.06	0.28	-0.35	0.50
1988	0.24	-0.10	-0.09	0.40	-0.02	0.15	0.32
1989	-0.84	-0.86	-0.64	-0.73	-0.78	-0.56	-0.36
1990	-0.70	-0.87	-0.72	-0.51	-0.58	-0.76	0.19
1991	-0.54	-0.86	-0.26	-0.45	0.11	-0.71	0.43
1992	0.22	0.04	0.27	0.19	0.29	0.13	0.46
1993	-0.11	0.12	-0.88	0.02	0.13	-0.17	0.47
1994	-0.38	-0.52	-0.78	-0.53	-0.37	-0.71	0.27
1995	0.05	-0.71	-0.79	-0.53	0.15	-0.18	0.54

Source: As indicated in table 1.

in five markets negative relationship between these two variables is found. These markets are: Jalna, Manwat, Latur, Udgir and Nanded. A positive relationship between arrivals and prices of greengram is found in case of Aurangabad, Partur, Parbhani and Kallam markets. Hence it is difficult to state the exact nature of relationship between arrivals and prices of greengram at the regulated markets during the period under consideration. Because the greengram is a short-duration crop and generally a large part of greengram production is sold in peak season in most of the markets.

Table 6

Coefficients of Correlation Between Monthly Prices and Arrivals of Greengram for Different Regulated Markets

	Markets				
Years	*Aurangabad*	*Jalna*	*Partur*	*Parbhani*	*Manwat*
1981	0.32	-0.49	0.75	0.24	0.28
1982	0.04	-0.70	-0.56	-0.61	-0.54
1983	-0.05	-0.78	-0.49	-0.54	-0.43
1984	0.27	0.44	0.46	0.34	0.35
1985	0.12	-0.50	0.04	-0.29	-0.11
1986	0.18	-0.21	0.11	0.32	0.10
1987	-0.49	-0.49	-0.47	0.51	-0.45
1988	-0.12	0.04	-0.03	-0.66	-0.18
1989	0.08	0.25	0.01	0.56	0.38
1990	-0.48	-0.45	0.06	0.24	-0.63
1991	0.33	0.30	0.45	0.63	0.38
1992	-0.37	-0.55	0.13	-0.55	-0.08
1993	0.19	-0.16	-0.43	-0.45	-0.77
1994	0.46	0.23	0.22	0.29	0.18
1995	0.19	0.27	0.35	0.09	0.31

	Markets			
Years	*Latur*	*Udgir*	*Kallam*	*Nanded*
1981	-0.04	-0.23	0.10	-0.34
1982	0.10	0.07	0.004	-0.73
1983	-0.07	0.11	0.28	-0.46
1984	-0.43	0.39	0.22	-0.58
1985	-0.49	-0.09	0.04	-0.32
1986	0.23	-0.37	0.70	-0.16
1987	-0.07	-0.31	0.27	-0.32
1988	-0.66	-0.32	-0.40	-0.92
1989	0.34	0.35	-0.13	-0.06
1990	-0.65	-0.08	0.17	-0.12
1991	0.42	0.42	0.30	0.45
1992	-0.33	-0.68	0.17	-0.08
1993	-0.36	-0.35	0.09	-0.34
1994	0.82	0.36	0.19	0.14
1995	0.08	0.35	0.35	0.46

Source: As indicated in table 1.

Safflower

Correlation coefficients presented in table 7 show that there has been a very weak relationship between arrivals and prices of safflower at the selected regulated markets. Out of 11 markets considered there has been a clear cut negative relationship between arrivals and prices

Table 7
Coefficients of Correlation Between Monthly Prices and Arrivals of Safflower for Different Regulated Markets

	Markets					
Years	*Aurangabad*	*Lasur*	*Jalna*	*Partur*	*Parbhani*	*Anwat*
1981	-0.60	-0.47	-0.54	-0.59	-0.55	-0.70
1982	-0.76	-0.71	-0.78	-0.93	-0.71	-0.74
1983	-0.09	-0.52	-0.76	-0.08	-0.28	-0.76
1984	-0.40	-0.26	-0.46	0.09	-0.33	-0.13
1985	-0.65	-0.17	-0.46	-0.08	-0.14	-0.24
1986	-0.66	-0.67	-0.87	-0.58	-0.70	-0.52
1987	-0.63	-0.09	-0.66	0.13	-0.35	-0.40
1988	-0.39	0.32	-0.20	0.13	-0.09	0.32
1989	-0.28	-0.30	-0.33	-0.39	-0.47	-0.40
1990	-0.50	-0.45	-0.67	-0.03	-0.72	-0.43
1991	-0.51	-0.39	-0.62	0.05	0.26	0.26
1992	-0.22	-0.15	0.14	0.26	0.21	0.37
1993	-0.39	-0.50	-0.83	0.28	-0.07	0.25
1994	-0.46	0.19	0.06	0.08	0.24	0.17
1995	-0.78	-0.68	0.08	0.44	0.44	0.02

	Markets				
Years	*Beed*	*Manjlegaon*	*Latur*	*Osmanabad*	*Kallam*
1981	0.01	-0.46	-0.52	-0.54	-0.39
1982	-0.18	-0.55	-0.77	-0.86	-0.41
1983	0.25	-0.71	-0.63	-0.61	-0.68
1984	-0.63	-0.02	-0.44	-0.13	-0.15
1985	0.37	-0.20	-0.61	0.21	-0.27
1986	0.04	-0.68	-0.62	-0.54	-0.67
1987	-0.23	-0.22	-0.52	-0.48	-0.51
1988	-0.02	0.41	-0.24	-0.04	-0.22
1989	0.19	-0.35	-0.39	-0.46	-0.52
1990	-0.10	0.61	-0.52	-0.49	-0.52
1991	-0.03	-0.42	-0.69	0.61	-0.21
1992	0.90	0.18	-0.13	0.62	0.16
1993	0.85	-0.05	-0.61	0.15	-0.64
1994	0.56	0.54	-0.36	0.19	-0.44
1995	-0.01	0.13	-0.69	0.07	-0.74

Source: As indicated in table 1.

of safflower during all years in case of Aurangabad and Latur markets. In case of Lasur, Jalna, Parbhani, Manwat, Manjlegaon, Osmanabad, Kallam and Beed markets there has been a negative relationship between arrivals and prices of safflower during most of the years. Partur is the only one market where positive relationship between arrivals and prices of this commodity is found in eight years out of 15 years. In case of this commodity it can broadly by concluded that there has been a negative relationship between its arrivals and prices in most of the markets and during most of the years.

Sunflower

In case of sunflower the results shown in table 8 indicate that out of six markets considered in three markets (viz., Latur, Osmanabad and Kallam) there has been a negative relationship between arrivals and prices. In the remaining three markets (viz., Parbhani, Beed and Manjlegaon) a positive relationship is found. From this it is difficult to state the exact nature of relationship between arrivals and prices of sunflower at regulated markets in Marathwada during the period of this study.

Groundnut

In case of groundnut a negative relationship between arrivals and prices is found in case of four markets viz., Manjlegaon, Latur, Osmanabad and Kallam. And in the remaining markets a positive relationship between arrivals and prices of this commodity is found. The markets where positive relationship appeared are Jalna, Manwat, Parbhani, Beed, Udgir and Nanded. It is difficult to identify the exact nature of the relationship between arrivals and prices of groundnut because groundnut is a cash crop grown in two seasons.

SECTION B : SEASONAL PATTERN OF MARKET ARRIVALS AND PRICES

This section deals with the seasonal pattern of market arrivals and prices of foodgrains and oilseeds at regulated markets in Marathwada. The commodity-wise analysis of the seasonal pattern of market arrivals and prices is presented below :

Table 8
Coefficients of Correlation Between Monthly Prices and Arrivals of Sunflower For Different Regulated Markets

			Markets			
Years	*Parbhani*	*Beed*	*Manjlegaon*	*Latur*	*Osmanabad*	*Kallam*
1981	0.21	0.04	0.24	-0.53	-0.61	-0.79
1982	0.33	0.04	0.52	0.11	0.26	0.36
1983	0.29	0.11	0.56	-0.45	-0.31	-0.15
1984	0.19	-0.36	-0.21	-0.60	-0.28	-0.29
1985	0.60	0.22	0.35	-0.71	-0.10	-0.50
1986	-0.57	0.34	0.21	-0.65	-0.08	0.12
1987	0.13	0.06	-0.02	-0.01	-0.58	-0.56
1988	-0.38	-0.33	0.33	0.18	0.15	0.09
1989	0.10	0.32	-0.46	0.04	-0.58	-0.51
1990	0.50	0.43	0.20	-0.48	-0.21	-0.38
1991	0.14	0.01	-0.55	-0.64	0.002	-0.39
1992	-0.34	-0.63	-0.70	-0.45	-0.55	-0.52
1993	-0.15	0.04	-0.01	-0.56	-0.10	-0.76
1994	0.45	0.47	0.21	-0.36	-0.20	-0.71
1995	0.29	-0.48	0.29	0.39	-0.34	-0.13

Source: As indicated in table 1.

Wheat

The table 10 shows that in case of wheat about 44 to 68 per cent of the total arrivals appeared on the market in peak season, at lowest prices. Among the markets the highest percentage of arrivals of wheat in peak season is found in Naigaon (68 per cent), followed by Manwat (64 per cent), Manjlegaon (60 per cent), Udgir (59 per cent), Latur (58 per cent), Kallam (56 per cent), Partur (56 per cent) and Osmanabad (55 per cent) markets and the lowest percentage in case of Nanded (44 per cent), Lasur (45 per cent) and Aurangabad (50 per cent) markets. The prices of wheat in peak season are found to be lower as compared to that in mid and lean seasons because of higher arrivals . This indicates that the larger arrivals of wheat during peak season push the prices of wheat to downward direction.

On the contrary, 13 to 24 per cent of the total arrivals of wheat entered into the market during lean season, when the average price was the maximum. Among the markets, the highest percentage of arrivals of wheat in lean season is found in case of Lasur and Parbhani (24 per cent), followed by Aurangabad (21 per cent), Jalna,

Table 9
Coefficients of Correlation Between Monthly Prices and Arrivals of Groundnut for Different Regulated Markets

			Markets		
Years	*Jalna*	*Parbhani*	*Manwat*	*Beed*	*Manjlegaon*
1981	0.51	0.22	0.46	0.09	-0.65
1982	0.22	0.05	0.39	-0.19	-0.49
1983	0.22	0.11	0.13	0.14	-0.27
1984	-0.42	0.51	-0.60	-0.05	0.06
1985	0.18	0.35	-0.57	-0.03	-0.19
1986	0.14	-0.59	0.38	0.37	-0.05
1987	0.48	0.41	0.41	0.19	-0.37
1988	-0.58	0.54	0.18	-0.52	-0.56
1989	0.48	0.41	0.40	0.30	0.56
1990	0.49	0.03	0.06	0.74	-0.03
1991	-0.14	-0.11	0.10	-0.17	0.16
1992	0.25	0.24	-0.29	-0.50	-0.74
1993	0.18	0.43	0.15	0.13	0.23
1994	0.30	0.43	0.53	0.40	-0.41
1995	0.38	0.53	0.22	-0.03	-0.38

			Markets		
Years	*Latur*	*Udgir*	*Osmanabad*	*Kallam*	*Nanded*
1981	0.20	-0.02	-0.61	0.34	0.11
1982	0.04	-0.09	-0.83	-0.60	0.11
1983	-0.39	0.18	-0.54	-0.56	0.06
1984	-0.11	-0.30	-0.40	-0.05	0.34
1985	-0.31	0.24	0.14	-0.19	0.15
1986	-0.74	0.84	0.35	0.65	0.25
1987	-0.19	0.32	-0.57	-0.09	0.40
1988	0.66	0.03	-0.35	-0.51	0.62
1989	0.09	0.60	0.39	0.19	-0.31
1990	-0.08	0.44	-0.05	-0.12	0.45
1991	-0.41	0.11	0.54	-0.40	-0.25
1992	-0.28	0.15	-0.57	-0.40	-0.18
1993	-0.02	0.32	0.04	0.10	-0.09
1994	0.02	0.31	-0.36	-0.53	0.35
1995	-0.52	0.57	-0.53	-0.40	0.21

Source: As indicated in table 1.

Table 10
Seasonal Pattern of Arrivals and Prices of Wheat at Regulated Markets During 1981-95

(Arrivals in Quintals)
(Price in Rs. per Quintal)

Seasons	AURANGABAD Arrivals	AURANGABAD Price	LASUR Arrivals	LASUR Price	JALNA Arrivals	JALNA Price	PARTUR Arrivals	PARTUR Price	PARBHANI Arrivals	PARBHANI Price	MANWAT Arrivals	MANWAT Price	BEED Arrivals	BEED Price
Peak	16168.53 (49.75)	315.86	8508.60 (45.50)	318.46	25582.99 (54.57)	329.80	3053.68 (56.13)	319.71	3599.94 (52.74)	330.24	3833.40 (64.45)	322.68	8656.20 (53.84)	344.70
Mid	9403.53 (28.94)	322.86	5698.40 (30.47)	326.23	11808.20 (25.19)	334.53	1463.54 (26.91)	336.05	1589.54 (23.28)	337.57	1324.99 (22.24)	338.61	4461.34 (27.74)	353.96
Lean	6922.86 (21.31)	347.20	4491.80 (24.03)	354 41	9486.33 (20,24)	367.83	922.94 (16.96)	361.90	637.13	352.06 (13.31)	789.34 (18.42)	362.58	2961.66 (23.98)	374.21
Total	32494.92 (100)		18698.80 (100)		46877.52 (100)		5440.16 (100)		5826.61 (100)		5947.73 (100)		16079.20 (100)	

Seasons	MANJLEGAON Arrivals	MANJLEGAON Price	LATUR Arrivals	LATUR Price	UDGIR Arrivals	UDGIR Price	OSMANABAD Arrivals	OSMANABAD Price	KALLAM Arrivals	KALLAM Price	NANDED Arrivals	NANDED Price	NAIGAON Arrivals	NAIGAON Price
Peak	3383.53 (59.90)	326.76	15581.87 (58.64)	340.48	2668.27 (59.26)	332.81	1649.86) (55.20	343.66	2700.07 (56.17)	329.80	9157.40 (44.73)	307.21	2640.06 (68.34)	281.67
Mid	1270.73 (22.50)	340.69	5628.47 (21.18)	347.36	314.85 (1259.40)	753.21	359.96 (25.20)		1243.79 (25.87)	347.88	7067.87 (34.53)	322.08	661.66 (17.13)	271.33
Lean	993.86 (17.60)	366.25	5363.67 (20.18)	365.50	574.99 (12.77)	333.85	585.80 (19.60)	375.85	863.13 (17.96)	353.23	4245.67 (20.74)	337.49	561.20 (14.53)	286.91
Total	5648.12 (100)		26574.01 (100)		4502.6 6(100)		2988.87 (100)		4806.99 (100)		20470.94		3862.92 (100)	

Peak Season = February-May; Mid-season = June-September; Lean season = October-January.
Source: Calculations are based on the data on monthly arrivals and prices compiled from the office records of regulated markets.

Latur and Nanded (20 per cent) markets and the lowest percentage of arrivals in case of Udgir and Manwat (13 per cent), followed by Naigaon (14 per cent) and Partur (17 per cent) markets.

Bajra

In case of bajra, we can see from table 11 that 44 to 93 per cent of the total arrivals appeared on the market during the peak season, when the price was the lowest. The proportion of market arrivals appeared in peak season was the highest in Manwat (93 per cent), followed by Partur (92 per cent) and Kallam (80 per cent) market and the lowest percentage of arrivals in Beed (44 per cent) and Latur (52 per cent) markets. In case of bajra the Latur market indicates highest price and Partur market shows lowest price in peak season. This is because demand for bajra in Partur market is lower than the Lasur market.

In case of bajra it seems that the percentage of market arrivals in peak season is very high. This can be attributed to the fact that the farmers have to sell their output soon after the harvest at low prices due to their money obligations and the lack of storage facilities. The data shown in table 11 shows that the proportion of arrivals of bajra in mid season is lower than that in peak season and higher than that in lean season but mid season prices are higher than peak season prices and lower than lean season prices, in case of most of the regulated markets.

The average price of bajra was the maximum during lean season. But the proportion of the total market arrivals of bajra appeared on the market in this season was as high as 19 per cent in Beed and 17 per cent in Aurangabad markets and as low as about one per cent in Partur and Manwat markets and three per cent in Kallam market.

Kharif Jowar

Table 12 shows that in case of Kharif jowar 43 to 92 per cent of the total arrivals appeared on the market in peak season at the lowest price. The proportion of arrivals of Kharif jowar in peak season is the highest in case of Partur (93 per cent) market, followed by Lasur (75 per cent) and Osmanabad (70 per cent) markets and the lowest in case of Udgir (43 per cent), Latur (56 per cent) and Kallam (58 per cent) markets.

Table 11
Seasonal Pattern of Arrivals and Prices of Bajra at Regulated Markets During 1981-95

(Arrivals in Quintals)
(Price in Rs. per Quintal)

Seasons	AURANGABAD Arrivals	AURANGABAD Price	LASUR Arrivals	LASUR Price	JALNA Arrivals	JALNA Price	PARTUR Arrivals	PARTUR Price	MANWAT Arrivals	MANWAT Price
Peak	20140.67 (61.89)	224.46	16130.40) (58.95	240.68	64799.40 (73.32)	221.46	4243.40 (91.99)	188.01	2646.46 (92.81)	209.03
Mid	6925.20 (21.28)	209.45	7978.07 (29.16)	250.76	16902.80 (19.12)	223.25	327.20 (7.09)	196.96	175.67 (6.16)	212.35
Lean	5478.94 (16.83)	243.70	3253.99 (11.89)	239.43	6681.60 (7.56)	228.05	42.40 (0.92)	163.53	29.27 (1.03)	132.51
Total	32544.81 (100)		27362.46 (100)		88383.80 (100)		4613.00 (100)		2851.40 (100)	

Seasons	BEED Arrivals	BEED Price	MANJLEGAON Arrivals	MANJLEGAON Price	LATUR Arrivals	LATUR Price	UDGIR Arrivals	UDGIR Price	KALLAM Arrivals	KALLAM Price
Peak	24825.87 (43.58)	207.76	13512.67 (67.24)	199.50	25460.80 (52.40)	212.66	2328.93 (67.23)	207.03	1450.87 (80.38)	203.40
Mid	21489.40 (37.72)	358.81	4979.34 (24.77)	228.00	16697.54 (34.36)	215.08	914.80 (26.41)	201.78	301.33 (16.69)	197.63
Lean	10655.07 (18.70)	284.95	1604.93 (7.99)	219.08	6431.19 (13.24)	221.01	220.40 (06.36)	186.78	52.87 (2.93)	184.26
Total	56970.34 (100)		20096.94 (100)		48589.53 (100)		3464.13 (100)		1805.07 (100)	

Peak Season = September-December; Mid-season = January-April; Lean season = May-August.
Source: As indicated in table 10.

Table 12
Seasonal Pattern of Arrivals and Prices of Kharif Jowar at Regulated Markets During 1981-95

(Arrivals in Quintals)
(Price in Rs. per Quintal)

Seasons	AURANGABAD		LASUR		JALNA		PARTUR		PARBHANI		BEED	
	Arrivals	Price	Arrivals	Price	Arrivals	Price	Arrivals	Price	Arrivals	Price	Arrivals	Price
Peak	18513.20	209.85	8013.33)	205.88	40172.33	191.00	5570.20	189.60	6459.67	177.05	4768.14	199.06
	(65.25)		(74.96		(63.82)			(92.89)		(68.83)		(66.01)
Mid	4998.79	193.51	1193.61	184.76	13106.87	196.98	329.61	166.75	2124.27	172.76	1576.93	205.53
	(17.62)		(11.16)		(20.82)		(5.50)		(22.64)		(21.83)	
Lean	4860.66	202.68	1484.07	164.48	9666.13	208.91	96.54	167.25	800.54	173.08	877.87	214.30
	(17.13)		(13.88)		(15.36)		(1.61)		(8.53)		(12.16)	
Total	28372.65		10691.01		62945.33		5996.35		9384.48		7222.94	
	(100)		(100)		(100)		(100)		(100)		(100)	

Seasons	LATUR		UDGIR		OSMANABAD		KALLAM		NAIGAON	
	Arrivals	Price	Arrivals	Price	Arrivals	Price	Arrivals	Price	Arrivals	Price
Peak	104765.54	191.88	25078.93	175.56	6888.40	188.80	13360.73	191.16	6267.73	166.86
	(55.91)		(43.18)		(70.36)		(58.51)		(57.12)	
Mid	50796.07	196.93	17270.60	177.66	1862.47	197.76	6517.47	196.33	3058.47	166.96
	(27 11)		(29.73)		(19.02)		(28.54)		(27.87)	
Lean	31828.47)	209.30	15733.07	184.51	1039.88	205.66	2958.39	216.76	1646.34	162.73
	(16.98		(27.09)		(10.62)		(12.95)		(15.01)	
Total	187390.08		58082.60		9790.75		22836.59		10972.54	
	(100)		(100)		(100)		(100)		(100)	

Peak season: October-January; Mid-season: February-May; Lean season : June-September.
Source: As indicated in table 10.

On the contrary, 2 per cent to 27 per cent of the total arrivals of Kharif jowar entered into the market during the lean season when the price was maximum. Among the markets the highest percentage of arrivals of this commodity in lean season is found in case of Udgir (27 per cent), Aurangabad (17 per cent) and Latur (17 per cent) markets and the lowest in case of Partur (2 per cent), Parbhani (8 per cent) and Osmanabad (11 per cent) markets.

The proportion of the total market arrivals of this commodity in mid-season is found to be relatively higher as compared to lean season but lower as compared to peak season. But the prices of Kharif jowar in mid season are higher than its peak season prices and lower than lean season prices.

Rabi Jowar

In case of rabi jowar the results shown in table 13 indicate that 23 to 57 per cent of the total arrivals of this commodity appeared on the market in peak season at the lowest price. The proportion of the total arrival in peak season is very high in some markets e.g. Manwat and Udgir (57 per cent). The lowest proportion of total arrivals of rabi jowar in peak season is found in case of Nanded (22 per cent) and Lasur (40 per cent) markets.

On the contrary, 16 to 32 per cent of the total arrivals of rabi jowar entered into the market during lean season when the prices were maximum. Among the markets the highest percentage of market arrivals of rabi jowar in lean season is found in case of Nanded (32 per cent) and Parbhani (27 per cent) markets and the lowest percentage in case of Jalna (17 per cent) and Manjlegaon (18 per cent) markets.

Tur

In case of tur, we can see from table 14 that 61 to 87 per cent of the total arrivals appeared on the market during the peak season when the prices were the lowest. Among the markets, the highest percentage of arrivals of this commodity in peak season is found in case of Aurangabad (87 per cent), followed by Partur (84 per cent), Beed (83 per cent) and Kallam (82 per cent) markets. After sugarcane and cotton, tur is grown as a cash crop in Marathwada and therefore this crop is marketed soon after the harvest due to

Table 13
Seasonal Pattern of Arrivals and Prices of Rabi Jowar at Regulated Markets During 1981-95

(Arrivals in Quintals)
(Price in Rs. per Quintal)

Seasons	AURANGABAD Arrivals	AURANGABAD Price	LASUR Arrivals	LASUR Price	JALNA Arrivals	JALNA Price	PARTUR Arrivals	PARTUR Price	PARBHANI Arrivals	PARBHANI Price	MANWAT Arrivals	MANWAT Price	BEED Arrivals	BEED Price
Peak	12360.67	304.71	15689.27	303.56	52874.13	298.81	7137.54	282.28	3530.00	300.66	8063.53	289.56	22510.60	298.45
	(48.65)		(39.72)		(55.37)		(53.61)		(42.17)		(57.28)		(48.78)	
Mid	7213.67	310.86	15693.99	317.66	25949.20	307.11	4183.87	292.68	2560.27	301.18	3664.54	298.81	15174.41	306.06
	(28.39)		(39.73)		(27.17)		(31.16)		(30.59)		(26.04)		(32.88)	
Lean	5832.80	323.41	8115.67	327.98	16675.27	320.10	2102.86	302.102	280.00	324.81	2345.80	305.78	8465.74	320.02
	(22.96)		(20.55)		(17.46)		(15.68)		(27.24)		(16.68)		(18.34)	
Total	25407.14		39498.93		95498.60		13424.27		8370.27		14073.87		46150.75	
	(100)		(100)		(100)		(100)		(100)		(100)		(100)	

Seasons	MANJLEGAON Arrivals	MANJLEGAON Price	LATUR Arrivals	LATUR Price	UDGIR Arrivals	UDGIR Price	OSMANABAD Arrivals	OSMANABAD Price	KALLAM Arrivals	KALLAM Price	NANDED Arrivals	NANDED Price
Peak	18214.87	300.11	8899.80	325.20	892.87	305.53	2565.46	305.38	2712.14	310.96	5959.86	212.03
	(52.49)		(45.04)		(57.02)		(45.33)		(53.04)		(22.47)	
Mid	10348.27	299.15	6328.79	323.26	362.80	265.98	1893.46	316.46	1596.60	312.45	12118.53	215.48
	(29.82)		(32.03)		(23.16)		(33.46)		(31.23)		(45.69)	
Lean	6140.81	307.46	4532.80	344.03	310.33	346.71	1200.00	328.30	804.40	319.11	8445.40	207.26
	(17.69)		(22.93)		(19.82)		(21.21)		(15.73)		(31.84)	
Total	34703.95		19761.39		1566.00		5658.92		5113.14		26523.79	
	(100)		(100)		(100)		(100)		(100)		(100)	

Peak season : March-June; Mid-season : July-October; Lean season : November-February
Source: As indicated in table 10.

Table 14

Seasonal Pattern of Arrivals and Prices of Tur at Regulated Markets During 1981-95

(Arrivals in Quintals)
(Price in Rs. per Quintal)

Seasons	AURANGABAD Arrivals	AURANGABAD Price	LASUR Arrivals	LASUR Price	JALNA Arrivals	JALNA Price	PARTUR Arrivals	PARTUR Price	PARBHANI Arrivals	PARBHANI Price	MANWAT Arrivals	MANWAT Price	BEED Arrivals	BEED Price
Peak	24678.52 (86.50)	690.43	9565.67 (80.86)	685.30	62046.46 (78.06)	702.86	14912.93 (84.40)	689.15	20590.93 (74.58)	674.28	20807.40 (76.49)	678.28	22621.06 (82.77)	696.51
Mid	2695.93 (9.45)	686.76	1437.01 (12.30)	697.82	14260.99 (17.95)	721.43	2616.07 (14.81)	735.96	6054.80 (21.93)	686.73	5419.34 (19.93)	708.60	4101.54 (15.01)	734.38
Lean	1154.53 (4.05)	657.93	682.14 (5.84)	636.58	3173.93 (3.99)	747.38	138.93 (0.79)	611.48	964.20 (3.49)	706.29	974.32 (3.58)	652.26	607.88 (2.22)	734.80
Total	28528.98 (100)		11684.82 (100)		79481.38 (100)		17667.93 (100)		27609.93 (100)		27201.06 (100)		27330.48	

Seasons	MANJLEGAON Arrivals	MANJLEGAON Price	LATUR Arrivals	LATUR Price	UDGIR Arrivals	UDGIR Price	OSMANABAD Arrivals	OSMANABAD Price	KALLAM Arrivals	KALLAM Price	NANDED Arrivals	NANDED Price	NAIGAON Arrivals	NAIGAON Price
Peak	20530.53 (76.48)	653.16	93106.74 (60.86)	720.83	42660.40 (66.66)	705.60	6485.47 (72.27)	683.31	6597.54 (82.04)	677.51	12288.34 (61.45)	674.94	1543.40 (72.90)	587.10
Mid	5915.33 (22.03)	725.16	50038.54 (32.71)	753.28	19384.40 (30.29)	754.85	2108.86 (23.50)	724.41	1320.80 (16.42)	711.76	6040.79 (30.20)	709.83	528.46 (24.96)	605.14
Lean	401.33 (1.49)	700.21	9835.14 (6.43)	780.13	1953.20 (3.05)	733.48	379.47 (4.23)	730.88	123.74 (1.54)	638.26	1670.66 (8.35)	747.54	45.34 (2.14)	322.01
Total	26847.19 (100)		152980.42 (100)		63998.00 (100)		8973.80 (100)		8042.08 (100)		19999.79 (100)		2117.20 (100)	

Peak season : December-March; Mid-season : April-July; Lean season : August-November.
Source: As indicated in table 10.

money obligations of the farmers. The data presented in table 14 reveals that one per cent to 8 per cent of the total arrivals of tur entered into the market during the lean season, when the prices were the highest.

Greengram

Table 15 shows that the proportion of the total arrivals of greengram appeared on the market during peak season was the highest in Partur (97 per cent) market when its price was the minimum. It was also very high in case of Manwat (95 per cent), Aurangabad (94 per cent), Udgir (92 per cent) and Jalna (92 per cent) markets. The tendency of the greengram growers is that they sell their entire output soon after the harvest of this crop after having left a very negligible part of output for their home consumption. Therefore, the proportion of the total market arrivals marketed during the mid and lean seasons is very negligible.

Safflower

In case of safflower it is seen from table 16 that 59 to 89 per cent of the total arrivals of this commodity appeared on the market during the peak season when the prices were at the lowest level. Among the markets, the highest percentage of the total arrivals of this commodity in peak season is found in case of Partur (89 per cent), followed by Manwat (86 per cent), Osmanabad (84 per cent) and Kallam (83 per cent) markets. In case of safflower peak season is characterized by higher arrivals and lower prices and the mid and lean seasons by lower arrivals and higher prices at all the markets. The proportion of the total arrivals of this commodity entered into the market ranged between 2 to 16 per cent.

Sunflower

In case of sunflower the proportion of the total arrivals appeared during peak season ranged between 48 to 58 per cent at different markets, when the prices were at the lowest level. Among the markets, the highest percentage of the total arrivals of sunflower is found in case of Beed (58 per cent), followed by Kallam (57 per cent), Parbhani (55 per cent), Osmanabad (49 per cent), Latur and Manjlegaon (48 per cent) markets.

Table 15

Seasonal Pattern of Arrivals and Prices of Greengram at Regulated Markets During 1981-95

(Arrivals in Quintals)
(Price in Rs. per Quintal)

Seasons	AURANGABAD Arrivals	AURANGABAD Price	JALNA Arrivals	JALNA Price	PARTUR Arrivals	PARTUR Price	PARBHANI Arrivals	PARBHANI Price	MANWAT Arrivals	MANWAT Price
Peak	8236.67 (94.28)	579.26	102855.60 (91.66)	608.90	11048.92 (96.87)	590.02	34577.73 (88.44)	595.18	16061.60 (94.70)	588.50
Mid	293.20 (3.36)	578.73	6955.20 (6.20)	608.56	317.66 (2.78)	517.03	4213.79 (10.78)	591.88	761.07 (4.49)	550.30
Lean	206.74 (2.36)	459.06	2399.27 (2.14)	638.18	39.79 (0.35)	424.63	304.06 (0.78)	544.25	137.87 (0.81)	488.95
Total	8736.61 (100)		112210.07 (100)		11406.37 (100)		39095.58 (100)		16960.54 (100)	

Seasons	LATUR Arrivals	LATUR Price	UDGIR Arrivals	UDGIR Price	KALLAM Arrivals	KALLAM Price	NANDED Arrivals	NANDED Price
Peak	26785.74 (77.15)	618.75	10144.14 (92.37)	597.98	465.60 (41.50)	537.59	4568.53 (67.38)	606.91
Mid	6390.07 (18.41)	609.01	614.67 (5.60)	614.10	619.66 (55.30)	564.38	1813.54 (26.75)	604.46
Lean	1541.66 (4.44)	626.88	223.13 (2.03)	579.69	35.19 (3.15)	438.76	398.33 (5.87)	599.35
Total	34717.47 (100)		10981.94 (100)		1120.45 (100)		6780.40 (100)	

Peak season : August-November; Mid-season : December-March; Lean season : April-July

Source: As indicated in table 10.

Table 16
Seasonal Pattern of Arrivals and Prices of Safflower at Regulated Markets During 1981-95

(Arrivals in Quintals)
(Price in Rs. per Quintal)

Seasons	*AURANGABAD* Arrivals	*AURANGABAD* Price	*LASUR* Arrivals	*LASUR* Price	*JALNA* Arrivals	*JALNA* Price	*PARTUR* Arrivals	*PARTUR* Price	*PARBHANI* Arrivals	*PARBHANI* Price	*MANWAT* Arrivals	*MANWAT* Price
Peak	7464.86 (74.42)	603.33	7004.34 (66.34)	600.61	20735.99 (58.24)	608.68	4451.20 (89.11)	594.76	5918.00 (79.33)	568.91	13574.26 (86.27)	599.26
Mid	1440.20 (14.36)	680.86	2369.13 (22.44)	697.90	8923.66 (25.06)	693.25	423.66 (8.48)	592.25	1073.73 (14.39)	575.81	1799.27 (11.44)	617.85
Lean	1125.40 (11.22)	701.86	1184.59 (11.22)	712.80	5944.60 (16.70)	706.53	120.40 (2.41)	585.33	469.54 (6.28)	609.41	360.93 (2.29)	518.57
Total	10030.46 (100)		10558.06 (100)		35604.25 (100)		4995.26 (100)		7460.27 (100)		15734.46 (100)	

Seasons	*BEED* Arrivals	*BEED* Price	*MANJLEGAON* Arrivals	*MANJLEGAON* Price	*LATUR* Arrivals	*LATUR* Price	*OSMANABAD* Arrivals	*OSMANABAD* Price	*KALLAM* Arrivals	*KALLAM* Price
Peak	7476.26 (70.93)	591.23	9252.13 (79.84)	611.43	6463.94 (66.61)	577.58	555.07 (84.16)	563.88	2022.80 (83.42)	574.65
Mid	2387.40 (22.65)	630.46	1708.88 (14.75)	681.23	1909.73 (19.68)	671.20	75.54 (11.45)	593.18	237.46 (9.79)	642.81
Lean	677.21 (6.42)	603.98	626.74 (5.41)	640.81	1330.73 (13.71)	667.93	28.93 (4.39)	511.73	164.66 (6.79)	628.25
Total	10540.87 (100)		11587.75 (100)		9704.40 (100)		659.54 (100)		2424.92 (100)	

Peak season : February-May; Mid-season : June-September; Lean season : October-January.
Source: As indicated in table 10.

It is also found that three to eleven per cent of the total arrivals of sunflower entered into the market in the lean season when the prices were maximum. Among the markets, the proportion of the total arrivals of this commodity in lean season is the lowest in Kallam (3 per cent), followed by Parbhani and Beed (10 per cent) markets, Manjlegaon, Latur and Osmanabad (11 per cent) markets

Groundnut

Groundnut is the most important oilseed crop grown in Marathwada. This crop is grown in kharif and summer seasons. During the study period, Jalna, Beed, Manjlegaon, Udgir and Osmanabad markets have received the kharif groundnut arrivals and the markets viz., Parbhani, Manwat, Latur, Kallam and Nanded have received the summer groundnut arrivals. Hence the peak, lean and mid seasons for these two groups of markets are different.

In case of the kharif groundnut the highest percentage of arrivals is found in case of Jalna (66 per cent), followed by Manjlegaon (63 per cent) and Osmanabad (50 per cent) markets when prices were at the minimum level.

In case of the summer groundnut the proportion of the total arrivals appeared in the market in the peak season is found to be the highest in case of Parbhani (89 per cent) followed by Nanded (86 per cent) market. In Latur (30 per cent) and Kallam (16 per cent) markets arrivals of groundnut are highest in lean season. But the lowest arrivals of groundnut is found in case of Parbhani (2 per cent) and Manwat (2 per cent) markets in lean season due to lack of storage facilities.

This study revealed that there has been a negative relationship between arrivals and prices of wheat in most of the markets during most of the years. In case of bajra the correlation coefficients between arrivals and prices are found to be positive for most of the markets. The correlation coefficients obtained for Kharif jowar have indicated a negative relationship in most of the markets. The relationship between arrivals and prices of rabi jowar is found to be negative but very weak in all the markets.

In case of tur, a highly negative relationship between arrivals and prices is found in most of the markets during the study period. There are few markets where arrivals and prices of tur have indicated a positive relationship. In case of greengram, the study does not indicate a clear cut nature of relationship between its arrivals and prices during the study period.

Arrivals and prices of safflower have indicated a negative relationship in most of the markets during the period under study. But as far as sunflower and groundnut are concerned, the relationship between arrivals and prices is found to be positive in some markets in some years and negative in other markets in other years.

Seasonal pattern of market arrivals has shown that the maximum portion of the total arrivals of almost all commodities considered is marketed during the peak season at lower prices during the study period. In case of wheat about 44 to 68 per cent of the total arrivals appeared on the market in peak season at lowest prices in different markets in Marathwada. In peak season the percentage of market arrivals of this commodity was as high as 66 per cent in Naigaon market and as low as 44 per cent in Nanded market. The large arrivals of wheat in peak season seems to have pushed its prices to downward direction in all the markets during the study period. In case of bajra, 44 to 93 per cent of the total arrivals are found to have appeared on the market in peak season. Further, this percentage is found to be very high in case of Manwat, Partur and Kallam markets. In case of bajra, the proportion of the total market arrivals is found to be very small during both mid and lean seasons. It means most part of the production of bajra enters into the market soonafter the harvest of this crop. The proportion of market arrivals of Kharif jowar in peak season was as high as 92 per cent in Partur market and as low as 43 per cent in Udgir market. In mid season the proportion of arrivals of this commodity is lower than that in peak season but higher than that in lean season. In case of rabi jowar, the percentage of total market arrivals in peak season is comparatively lower than that of other foodgrains in all the selected markets during the study period.

This study has indicated that the proportion of total market arrivals of pulses appeared in peak season was very high in all the markets except Latur and Kallam, when the prices were at lower level. The proportion of total arrivals of tur appeared on the market in peak season is found to be the highest in case of Aurangabad (86 per cent) market and the lowest for Latur (60 per cent) market.

Table 17
Seasonal Pattern of Arrivals and Prices of Sunflower at Regulated Markets During 1981-95

(Arrivals in Quintals)
(Price in Rs. per Quintal)

Seasons	MARKETS: PARBHANI Arrivals	PARBHANI Price	BEED Arrivals	BEED Price	MANJLEGAON Arrivals	MANJLEGAON Price	LATUR Arrivals	LATUR Price	OSMANABAD Arrivals	OSMANABAD Price	KALLAM Arrivals	KALLAM Price
Peak	3514.06 (54.94)	708.06	12024.07 (57.69)	767.41	9587.00 (47.74)	688.90	40358.93 (48.12)	752.25	2955.06 (49.29)	724.50	6411.14 (57.16)	722.41
Mid	2260.33 (35.33)	671.36	6749.60 (32.38)	716.35	8190.00 (40.78)	703.26	33906.87 (40.43)	736.33	2378.93 (39.68)	721.91	4477.28 (39.92)	710.79
Lean	622.53 (9.73)	676.70	2068.53 (9.93)	740.59	2305.34 (11.48)	722.63	9605.47 (11.45)	847.85	661.53 (11.03)	832.88	327.87 (2.92)	816.58
Total	6396.92 (100)		20842.20 (100)		20082.34 (100)		83871.27 (100)		5995.52 (100)		11216.29 (100)	

Peak season : October-January; Mid-season : February-May; Lean season : June-September
Source: As indicated in table 10.

Table 18
Seasonal Pattern of Arrivals and Prices of Groundnut at Regulated Markets During 1981-95

(Arrivals in Quintals)
(Price in Rs. per Quintal)

SEASONS	JALNA		BEED		MARKETS MANJLEGAON		UDGIR		OSMANABAD	
	Arrivals	*Prices*	*Arrivals*	*Prices*	*Arrivals*	*Prices*	*Arrivals*	*Prices*	*Arrivals*	*Prices*
Peak	3242.80 (65.76)	675.83	9453.27 (51.47)	736.63	14809.50 (62.71)	695.60	315.14 (37.05)	662.98	1425.99 (50.36)	687.05
Mid	951.14 (19.29)	539.23	7191.20 (39.15)	746.38	5314.53) (22.50	725.02	322.00 (37.86)	613.00	889.00 (31.39)	733.90
Lean	737.40 (14.95)	587.6	1722.66 (9.38)	763.52	3492.67 (14.79)	723.27	213.33 (25.09)	675.27	516.86 (18.25)	750.04
Total	4931.34 (100)		18367.13 (100)		23616.70 (100)		850.47 (100)		2831.85 (100)	

SEASONS	PARBHANI		MANWAT		MARKETS LATUR		KALLAM		NANDED	
	Arrivals	*Prices*	*Arrivals*	*Prices*	*Arrivals*	*Prices*	*Arrivals*	*Prices*	*Arrivals*	*Prices*
Peak	6417.54 (88.91)	722.75	3325.13 (42.00)	731.50	17841.40 (36.71)	806.93	1231.80 (40.36)	698.13	24272.47 (86.11)	813.63
Mid	641.34 (08.89)	656.72	4427.33 (55.92)	691.75	16099.14 (33.13)	745.56	1332.33 (43.66)	714.80	2977.53 (10.56)	778.58
Lean	159.00 (2.20)	523.76	165.39 (2.08)	555.30	14658.33 (30.16)	762.14	487.73 (15.98)	741.53	939.00 (3.33)	593.76
Total	7217.88 (100)		7917.85 (100)		48598.87 (100)		3051.86 (100)		28189.00 (100)	

Peak season : May-August; Mid-season : September-December; Lean season : January-April.
Source: As indicated in table 10.

In this study it is clearly found that the proportion of total market arrivals appeared on the market was as high as 89 per cent in Partur market and as low as 58 per cent in Jalna market in case of safflower, 57 per cent in Beed market and 48 per cent in Manjlegaon market in case of sunflower, and about 89 per cent in Parbhani market and 37 per cent in Latur and Udgir markets in case of groundnut. Further, this study has indicated that the prices of groundnut have been relatively lower in peak season than that in mid and lean seasons.

7

FINDINGS AND POLICY MEASURES

Agricultural prices play a significant role in national economy. National economy of India is still dominated by agriculture, because agriculture commands a substantial part of the total national income. There is a close relation between the agricultural sector and the whole national economy. Prices of agricultural products and particularly the foodgrain prices play an important role, not only in the agricultural sector but in the whole national economy of India. Agricultural prices affect the production of food and non-foodgrains and change the terms of trade between agriculture and non-agricultural sector. They also affect the entire cost structure and pattern of income distribution. Government revenue also gets affected from agricultural prices. Agricultural prices are important in determining the size of marketed surplus of agricultural products and foreign exchange earnings. Agricultural prices may play a more important role in signaling the government the need for greater investment in roads, power supply, extension, research and parts of infrastructure of technological change in agriculture.

Agricultural prices is a big problem in Indian agricultural sector, because agricultural production depends upon monsoons. Therefore the agricultural prices are subject to fluctuations any time. There are two aspects of agricultural prices problem. The first is : Are prices of agricultural products rational and incentives? The second aspect of agricultural prices problem is arbitrary variation in the prices of agricultural products. This research work is concerned with the second problem. In the free market economy the prices of agricultural commodities are unexpected and fluctuate violently, which have adverse effects on the pattern and volume of agricultural production. There are two types of price variation: spatial price variation and temporal price variation. Temporal price variations are further classified as short-time price variations, seasonal price variations, annual price variations, cyclical price variations, secular trend in prices and irregular price variations.

Short-time price variations are caused mainly by three factors viz., variations in receipts, temporary fluctuations in consumer demand and the experimentation involved in the discovery of supply and demand conditions and prices. The seasonal nature of agricultural production, low capacity of the small and marginal farmers to hold stock and to obtain credit to postpone their sales, less storability of agricultural products are some of the factors which explain the variations in arrivals and prices of agricultural commodities from one season to another within a year. Annual price variations result in fluctuations in the farmers income. If the demand for farm products is inelastic, the annual variation in yields cause prices to fluctuate. It means that when demand for farm products is inelastic, the producers incomes will vary from year to year with crop yields. A trend in prices is usually established on the basis of at least 10 to 15 years data. The trend is not concerned with the movement in prices from one year to another but for a large number of years say ten or more.

Agricultural price variations is a conspicuous phenomenon in India. There are number of other factors which result in fluctuations in the agricultural prices viz., change in monetary policy, fiscal policy, tariff policy, future price trend, transportation, storage, credit and processing facilities, effective marketing system, economic and political events, lack of effective price policy, change in technology, credit policy, tax rates, nature of production, change in market arrivals, etc. But the variations in agricultural prices are mainly attributed to inelastic nature of demand for agricultural products, inelastic supply of factors of production, domination of nature, fixed nature of cost of production, combination of different products, seasonal sale of agricultural products, speculative demand of wholesale traders, vicious circle between prices and sales, etc.

Agricultural prices are more sensitive to change in imbalance between supply and demand than the prices of industrial products.

As a result, price variation takes place by dimension of time, space and form. It means prices fluctuate from time to time, from market to market and from commodity to commodity. There are various effects of price variation on the national economy such as on economic life of farmers, allocation of resources, production pattern, monetary policy, fiscal policy, capital formulation, etc. Violent fluctuations in agricultural prices affect the inter-sectoral terms of trade and depress the incentive of farmers. Variations in agricultural prices have adverse effect on farmers income level and their economic prosperity. Such variations promote black marketing and speculation and reduce regional integration of agricultural markets. It has severe impact on agricultural production, producer's share in consumer's rupee and on government budget.

The present study was undertaken mainly to study the seasonal, annual and long term variations in arrivals and prices of six foodgrains and three oilseeds at 14 regulated markets in Marathwada during the period 1981-95. In this study an attempt was also made to study the relationship between arrivals and prices of foodgrains and oilseeds and seasonal pattern of arrivals and prices.

The main findings of this study are discussed in the subsequent section.

FINDINGS

1. The study has revealed that arrivals of foodgrains and oilseeds are subject to seasonal variation. This study has indicated that the seasonal variation in arrivals of wheat were comparatively higher in Jalna and Naigaon markets and lower in Osmanabad and Lasur markets. In case of bajra the higher variation of arrivals is found in Jalna and Manwat markets and lower in Kallam and Beed markets. Kharif jowar arrivals have indicated a higher degree of variation in Latur and Partur markets and lower variation in Beed and Udgir markets. In case of rabi jowar, its arrivals have shown the highest variation in Udgir market and lowest variation in Parbhani market. In case of tur the degree of variation in arrivals is found to be greater in Lasur and Aurangabad markets and lesser in Naigaon and Nanded markets. As far as greengram is concerned, the larger variation in its arrivals is found in Jalna and Aurangabad markets and smaller

variation in Nanded and Kallam.markets. In case of safflower, its arrivals have varied to a greater extent in Manwat and Partur markets and to a smaller extent in Osmanabad and Jalna markets. Sunflower arrivals have shown the smaller variation in Manjlegaon market and greater variation in Latur and Parbhani markets. Seasonal variation in groundnut arrivals is found to be the highest in Nanded market and lowest in Udgir and Latur markets during the study period.

2. This study has indicated that the monthly prices of wheat and bajra have varied to a greater extent in Partur and Beed markets respectively but the prices of both the commodities have exhibited lowest variation in Latur market. Kharif jowar prices have indicated large degree of variation in Latur market and smaller degree of variation in Udgir market. Rabi jowar prices have shown the highest seasonal variation in Udgir market and lowest seasonal variation in Nanded market. Prices of tur have varied to a greater extent in Naigaon market and to a smaller extent in Jalna market. In case of greengram the seasonal variation of its prices was the highest in Aurangabad market and the lowest in Latur market. Safflower prices have shown the highest seasonal variation in Partur market and lowest in Manjlegaon market . On the other hand, the degree of seasonal price variation of sunflower was the highest in Parbhani market and lowest in Manjlegaon market. But the variation in groundnut prices was the highest in Jalna market and the lowest in Manjlegaon market.

3. This study has revealed that the indices of monthly arrivals of wheat were relatively higher in the months of March, April and May and relatively lower in the months of October, November and December in all the markets during the study period. In case of bajra indices of arrivals were relatively higher in the months of November, December and January and relatively lower in the months of March, April and May. But the seasonal indices of arrivals of kharif jowar were comparatively higher in the months of October, November and December and lower in the months of June, July, August and September. In case of rabi jowar, the seasonal indices of arrivals were relatively higher in the months of March, April and May and relatively lower in the months of

November, December and January in all the markets during the study period. This study has indicated that arrivals of tur reach to a peak level in the months of December, January and February. And the arrivals of this commodity reach a low level in the months of September, October and November in all the markets. The seasonal indices of greengram arrivals are found to be comparatively higher in the months of September and October and relatively lower in the months of April, May and June in all the markets. In case of safflower seasonal indices of arrivals are found to be relatively higher in the months of March, April and May and relatively lower in the months of November, December and January in all the markets except Aurangabad market. Seasonal arrival indices of sunflower were higher in the months of November and December and relatively lower in the months of June, July, August and September. The periods of relatively higher and lower indices of arrivals of groundnut are found to be different for different markets.

4. This study has indicated that the seasonal index of wheat prices is comparatively higher in the months of November, December and January and lower in the months of March, April and May. In case of bajra, its price indices are found to be relatively higher in the months of December, January and February and relatively lower in the months of September, October and November in most of the markets. But the kharif jowar price indices are found to be comparatively higher and lower in different months in different markets. Seasonal indices of rabi jowar are found to be relatively higher in the months of November, December and January in all the markets except Nanded market and relatively lower in the months of March, April and May in all the markets except Udgir and Nanded markets. The seasonal price indices of tur were relatively higher in the months of June, July and August and relatively lower in January, February and March. But the seasonal indices of greengram prices are found to be higher in the months of October, November and December and lower in the months of April, May, June and July. In case of safflower, seasonal indices of its prices are higher in the months of March, April and May and smaller in the months of November, December and January in all the markets except Aurangabad

market. The sunflower price indices are found to be relatively higher in the months of June, July and August and relatively lower in the months of November and December. The periods of relatively higher and lower indices of prices of groundnut are found to be different for different markets.

5. The analysis of annual variation in the arrivals of foodgrains and oilseeds revealed that in case of wheat the annual variations in arrivals were relatively higher in Latur and Udgir markets and relatively lower in Osmanabad and Manjlegaon markets. In case of bajra the study indicated that the annual variation in arrivals was relatively higher in Kallam market and relatively lower in Aurangabad market. Kharif jowar arrivals have exhibited the highest degree of annual variation in Lasur and Beed markets and lowest degree of variation in Latur and Partur markets. Annual arrivals of rabi jowar have been subject to larger variation in Jalna and Udgir markets and smaller variation in Kallam market during the study period. But the annual variation of tur arrivals is comparatively higher in Latur and Aurangabad markets and lower in Naigaon and Parbhani markets during the study period. The degree of annual variation in arrivals of greengram was greater in Jalna and Aurangabad markets and smaller in Kallam and Partur markets. In case of safflower, annual variation in arrivals was the highest in case of Osmanabad market and the lowest in case of Latur market. The degree of annual variation in sunflower arrivals is found to be greater in Latur and Parbhani markets and smaller in Osmanabad and Kallam markets. In case of groundnut the degree of variation in its annual arrivals is found to be the highest in Nanded and Manwat markets and the lowest in Udgir and Latur markets during the study period.

6. The analysis of annual variation in the prices of foodgrains and oilseeds revealed that in case of wheat the annual variation in prices was comparatively higher in Partur market and lower in Naigaon market. Annual prices of bajra showed the highest variation in Lasur market and the lowest in Udgir market. Prices of Kharif jowar have varied to a larger extent in Beed and Aurangabad markets and to a smaller extent in Partur market. But in case of rabi jowar the annual variation in its prices is found to be the highest in Lasur market and the lowest in Nanded

and Parbhani markets. The degree of annual variation in tur prices was the highest in Latur and Lasur markets and lowest in Naigaon market. Annual prices of greengram have exhibited larger variation in Udgir market and smaller variation in Parbhani market. In case of safflower annual variation of price is found to be relatively higher in Latur market and smaller in Manwat market. This study has indicated that the degree of annual variation in prices of sunflower was the highest in Beed market and the lowest in Osmanabad market. Annual prices of groundnut have varied to a greater extent in Nanded market and to a smaller extent in Jalna market during the study period.

7. This study has indicated that the arrivals of wheat have not shown the similar trend in all the markets during the study period. The arrivals of wheat have shown a rising trend in seven markets and a decreasing trend in another seven markets. The annual rate of increase or decrease in arrivals is found to be different for different markets. In case of bajra, its arrivals have shown a rising trend in seven markets. However, the arrivals of bajra have shown a decreasing trend in three markets. Arrivals of bajra have shown the highest annual increase in Jalna market. Annual trend values of arrivals of this commodity have remained stagnant in Lasur market over the period under study. As the bajra crop requires less rainfall and unfortunately the rains in the recent years are poor, farmers have started growing more bajra. So, the bajra arrivals in these markets have showed an increase. Rabi jowar arrivals have shown a decreasing trend in nine markets. The decreasing trend is observed due to the reduced production of the crop and poor weather conditions, and an increasing trend in only four markets. The highest annual rate of decrease in arrivals of rabi jowar is found for Manjlegaon market and the lowest rate of decrease for Kallam market. Further, the highest rate of increase in arrivals of rabi jowar is found for Jalna market and the lowest rate of increase for Osmanabad market. In this study it is found that the arrivals of kharif jowar have shown a decreasing trend in all the markets except Latur and Naigaon. This may be due to fall in acreage under this crop in these market areas. The arrivals of tur have shown a rising trend in nine markets and a decreasing trend in

five markets during the study period. Its arrivals have recorded the highest annual rate of increase in Aurangabad market and the lowest rate of increase in Manwat market. Due to better production of the crop and attractive prices in the market, has resulted in increased arrivals. Greengram arrivals have shown a decreasing trend in six markets and an increasing trend in three markets. The annual rate of increase in greengram arrivals is found to be the highest in Jalna market and the lowest in Kallam market. This study has indicated that the arrivals of safflower have shown a decreasing trend in all the markets during the study period. Due to insufficient monsoons, the acreage under cultivation of this crop gets reduced. The highest annual rate of decrease in its arrivals is found for Jalna market and lowest for Osmanabad market. But sunflower arrivals have shown a positive trend in all the markets during the study period. The annual rate of increase in its arrivals is found to be the highest for Latur market and lowest for Osmanabad market. Groundnut arrivals have shown a rising trend in most of the markets (i.e. in seven out of 10 markets). Groundnut arrivals have exhibited the highest annual rate of increase in Beed market and the lowest rate of increase in Udgir market.

8. This study has indicated that the prices of all foodgrains and oilseeds have shown the rising trend in all the markets during the study period. This may be attributed to the factors such as population growth, changes in the weather conditions, increase in per capita income of the consumers and rising cost of inputs in farming. The prices of wheat have exhibited the highest annual rate of increase in Latur market and the lowest rate in Naigaon market. Annual increase in the prices of bajra is found to be the highest for Latur market and the lowest for Udgir market. The prices of rabi jowar have shown the highest annual rate of increase in Latur market and the lowest rate in Nanded market. The annual rate of increase in prices of kharif jowar has differed from market to market. The annual rate of increase in tur prices is found to be the highest for Latur market and the lowest for Naigaon market. Greengram prices have shown the highest annual rate of increase in Udgir market and the lowest rate of increase in Partur market. But in case of safflower prices the

highest rate of increase is found for Latur market and the lowest rate of increase for Parbhani market. Sunflower prices have increased at the highest annual rate in Beed market and at the lowest rate in Osmanabad market. Groundnut prices have increased at the highest annual rate in Nanded market and at the lowest rate in Parbhani market.

9. This study has revealed that there has been a negative relationship between arrivals and prices of wheat in most of the markets during most of the years. In case of bajra, the correlation coefficients between its arrivals and prices are found to be positive for most of the markets. The correlation coefficients obtained for kharif jowar have indicated a negative relationship between arrivals and prices in most of the markets. The relationship between arrivals and prices of rabi jowar is found to be negative but very weak in all the markets. In case of tur a highly negative relationship between its arrivals and prices is found in most of the markets during the study period. There are few markets where arrivals and prices of tur have indicated a positive relationship. This study does not indicate a clear cut nature of relationship between arrivals and prices of greengram during the study period. Arrivals and prices of safflower have indicated a negative relationship in most of the markets. The relationship between arrivals and prices of sunflower and groundnut is found to be positive in some markets in some years and negative in other markets in other years.

10. The study of seasonal pattern of market arrivals has shown that the maximum portion of total arrivals of almost all the commodities considered is marketed during the peak season at lower prices in all the markets in Marathwada during the study period. In case of wheat, 44 to 68 per cent of the total arrivals are found to have appeared on the market in peak season at lowest prices in different markets in Marathwada. The larger arrivals of wheat in peak season are found to have pushed its prices to downward direction in all the markets during the study period. In case of bajra, 44 to 93 per cent of the total arrivals are found to have appeared on the market in peak season during the study period. Further, this per centage is found to be very high in case of Manwat, Partur and Kallam markets. The

proportion of market arrivals of kharif jowar in peak season was as high as 92 per cent in Partur market and as low as 43 per cent in Udgir market.

This study has indicated that the proportion of total market arrivals of pulses (i.e. tur and greengram) appeared in peak season was very high in all the markets except Latur and Kallam, when the prices were minimum. The proportion of the total arrivals of tur appeared on the market in peak season was the highest in case of Aurangabad market and the lowest in case of Latur market. In this study it is found that the proportion of total market arrivals appeared on the market was as high as 89 per cent (in Partur market) and as low as 58 per cent (in Jalna market) in case of safflower, 57 per cent (in Beed) and 47 per cent (in Manjlegaon) in case of sunflower and about 89 per cent (in Parbhani market) and 37 per cent in (Latur and Udgir markets) in case of groundnut.

POLICY MEASURES

Agricultural prices play a very significant role in rural economy as well as in the whole national economy. The price variations have serious adverse effects on the economic life of the farmers. Considering the importance of agriculture and agricultural prices it is most essential to solve the problem of violent variation in prices of agricultural commodities and problems related to agricultural marketing. It is necessary to improve the economic position of farmers and induce them to enhance agricultural production by adopting new technology. The following policy measures are suggested for reducing the variation in agricultural prices and improving the structure of agricultural markets.

1. Economic poverty, which generates heavy burden of loan is a key problem of farmers and this problem is related with the variation in agricultural prices. Due to low money income most of the farmers borrow from the money lenders, which is an easy source of loan. Because of the pressure of repayment of loans taken from money lenders farmers sell their products to money lenders at low prices or in the market when prices are lower due to glut of supply in the peak season. So it is essential to increase the waiting power of

the farmers by providing credit at rational rate of interest. It is possible through extending services of commercial banks and cooperative banks to rural areas. A large number of farmers' cooperative credit societies are working in rural areas to meet the credit needs of farmers at cheaper rate. But the problem is that the functioning of these cooperative societies is not satisfactory and supply of credit by them is also not adequate to meet the farmers credit requirements. So the attention needs to be paid towards improving the functioning of such cooperative societies. The adequate finance should be provided to farmers through cooperative credit societies. Further, it is necessary that the flow of credit from these societies should be reached to small and marginal farmers.

2. Lack of storage facilities is an important problem in rural areas. It results in an increase in supply of agricultural products in post-harvest season which causes the prices to fall down. To avoid the glut of market arrivals in post-harvest season and consequent economic loss to farmers due to fall in prices, provision of storage facilities in rural areas is essential. Government should provide the storage facilities to farmers at cheap rates by constructing godowns in rural and market areas.

3. Government should enter into the process of price fixation at every stage for all commodities covering every sector of the economy. There is a need of enlarged and effective machinery to keep prices parity for various products in the agricultural sector, as well as between the agricultural and manufactured products. From this point of view, the present participation of government in price fixation is very limited and ineffective. As a result imbalance may emerge in the production of various agricultural commodities and between the various sectors of the economy.

4. It is much beneficial for small farmers to provide transport services through setting up of cooperative societies in rural areas. Any step taken to avoid price variations of agricultural products must start from village level. So the better way is

to develop strong and healthy cooperative movement at village level. The Government should concentrate on linking villages to towns through roads, so that the farmers would be able to sell their products in big markets where prices are generally higher.

5. Agricultural produce market committees can play an important role in marketing of agricultural products. Some important services may be provided to the farmers through such market committees. So the responsibilities of regulated market committees should be enlarged and they may be assigned the work of providing storage and transport facilities to farmers of the surrounding area. This policy would be more beneficial to small and marginal farmers.

6. The participation of Government as a buyer in agricultural markets is essential. But government should enter into agricultural markets as a competitor to private traders and not for monopoly procurement. Government should operate as a trader to purchase agricultural products mainly foodgrains in open market at the support prices which are rational and related to cost of cultivation. In a free market economy, the prices of agricultural products may not be rational and stable. On the other side, in case of monopoly procurement there are a number of defects in the implementation of this scheme. So to avoid the problems of free market and monopoly procurement, government should enter into the market as a purchase party like trader and should purchase the farm products at guaranteed prices. Since the existing Government machinery for procurement of agricultural products is not working properly and is not adequate, there is need of improving the existing machinery and setting up of an effective and extensive purchasing machinery in rural areas.

7. Most of the agricultural markets in India is in the private sector. In this sector, there is more possibility of creation of artificial scarcity, speculation and other malpractices to obtain the maximum monetary gain. Such practices result in erratic fluctuations in agricultural prices. So government should

have control on the purchase and sale of agricultural products by private traders. The government can exercise this control by adopting several measures such as issuing of the licences, declaration of stock by traders, statutory limit on stock, maintenance of proper accounts of their purchase and sale transactions and its submission periodically, etc. Existing rules for controlling the transactions of private traders need to be strictly and effectively implemented.

8. Government should go for the policy of "Agriculturalization of industry". It means that different types of processing units need to be established in rural areas on the field of the farmers particularly on cooperative basis. It is a way of moderating the wide fluctuations in agricultural prices and assuring the high income to farmers.

SELECT BIBLIOGRAPHY

BOOKS

Acharya, S.S., *Agricultural Production, marketing and price policy in India (A study of Pulses)*, Mittal Publications, Delhi, 1988.

Acharya, S.S. and Agrawal, N.L., *Agricultural prices-Analysis and policy,* Oxford and IBH Publishing Co. Pvt. Ltd., New Delhi, 1994.

Acharya, S.S. and Agrawal, N.L., *Agricultural Marketing in India,* Oxford and IBH Publishing Co. Pvt. Ltd., New Delhi, 1996.

Agrawal, N.L., *Agricultural prices and marketing in India,* Mittal Publications, Delhi, 1986.

Batra, M.M., *Agricultural Production, Prices and Technology,* Allied Publishers Pvt. Ltd., New Delhi, 1978.

Chalam, K.S. and Sanjeeva Rao, Agricultural Price Policy in India, in *Agricultural Marketing Perspective and Issues,* (Edited), Arihant Publishers, Jaipur, 1992.

Cohen, R.L., *The Economics of Agriculture,* James Nishbet and Co. Ltd., Cambridge at the University Press, 1955.

Dandekar, V.M., *Food and Freedom,* Dharwad University, Dharwad, 1967.

George, P.V., *Price Behaviour in India*, University of Bombay, 1968.

Gulati Ashok, *Agricultural Price Policy in India,* Concept Publishing Company, New Delhi, 1987.

Indian Society of Agricultural Economics, *Seminar on Marketing of Agricultural Commodities,* Seminar Series-V, 1965.

Jagdish Prasad and Arbind Prasad, *Indian Agricultural Marketing,* Mittal Publications, New Delhi, 1995.

Jain, S.C., *Agricultural Policy in India,* Allied Publishers, Bombay, 1967.

Jha, B.V., *Agricultural Price Stabilization in India,* Short Publications, Modern Street, Calcutta, 1974.

John Moore, Johl, S.S. and Khusro, A.M., *Indian Foodgrain Marketing,* Kahlon, A.S. and Tyagi, D.S., *Agricultural Price Policy in India,* Allied Publishers Pvt. Ltd., New Delhi, 1983.

Kainth, G.S., *Foodgrain Marketing System in India,* Associated Publishing House, New Delhi.

Kalra, O.P., *Agricultural Price Policy in India,* Popular Prakashan, Bombay, 1973.

Kaur Rajbans, *Agricultural Price Policy in Economic Development,* Kalyani Publishers, Ludhiana, 1975.

Ministry of Food and Agriculture, Government of India, *Price Policy in India,* 1963.

Mynard Geofry, *Economic Development and the Price Level,* Macmillan and Co. Ltd., London, 1962.

Nadkarni, M.V., *Agricultural Prices and Development with Stability,* National Publication House, 23, Darya Ganj, Delhi, 1973.

National Council of Applied Economic Reseárch, *Structure and Behavior of Prices of Foodgrains,* New Delhi, 1969.

Prasad, S.V., *Agricultural Marketing in India,* Mittal Publication, Delhi, 1985.

Satish Chandra, *Agricultural Price Policy in India,* Chug Publication, Allahabad, 1985.

Singh, L.P. and Subhash Garg, *Agricultural Marketing Perspective and Issues,* Arihant Publishers, Jaipur, 1992.

Singh, L.S., *Agricultural Price Policy and Stabilization Measures in India,* Capital Publishing House, Delhi, 1983.

Singh Raj Kumar, *Agricultural Price Policy in India,* Printwell Publishers, Jaipur, 1990.

Sinha, S.P., *Indian Agriculture : Its Fluctuating Fortunes,* Kitab Mahal Pvt. Ltd., Allahabad, 1965.

Saxena, H.M., *Regulated Agricultural Markets,* Rawat Publications, Jaipur, 1992.

Thomes and Foote, *Agricultural Prices,* McGraw Hill Book Company, New York, 1952.

Warren and Trelogan, *Agricultural Market Prices,* John Wiley and Ests, New York, 1951.

JOURNALS AND PERIODICALS

Adur, K.G., Price Behaviour of Agricultural commodities, *Southern Economist*, Vol. XXIV, No. 2, 15 May, 1985.

Antani, K.L. and Acharya, S.S., The Change in Behaviour of Market Arrivals and Their Effect on Prices of Wheat and Gram in Two Markets of Rajasthan, *Agricultural Marketing*, Vol. XXI, No. 4, January, 1979.

Balwinder Singh and Sidhu, D.S., Pattern of Markets Arrivals and Prices of Foodgrains in Punjab, *Agricultural Marketing*, Vol. XIV, No. 4, January, 1972.

Babar, V.S. and Lohar, N.S., Trends in Arrivals and Prices of Jaggery in Sangli Regulated Market, *Indian Journal of Agricultural Marketing*, Vol. VIII, No. 1, January-July, 1994.

Bhalla, G.S. and Gurmail Singh, Recent Developments in Indian Agriculture, *Economic and Political Weekly*, Vol. XXXII, No. 13, March, 1997.

Bhatia, M.S. and Sukheja, A.L., Behaviour of Paddy and Rice Prices - A Case Study, *Agricultural Marketing*, Vol. XIV, No. 1, April, 1971.

Brahma Prakash and Sushila Srivastava, Analysis of Market Arrivals and Prices of gram in Uttar Pradesh, *Indian Journal of Agricultural Economics*, Vol. 48, No. 3, July-September, 1993.

Brahma Prakash and Sushila Srivastava, Pattern of Marketing Arrivals and Prices of Gram in Uttar Pradesh, *Indian Journal of Agricultural Marketing*, Vol. VIII, No. 1, January-July, 1994.

Chatha, I.S. and Sidhu, D.S., Marketing of Kharif Pulses in Punjab, *Agricultural Marketing*, Vol. XXVI, January, 1984.

Chaturvedi, J.N., Prices and Price - Arrivals Response of Paddy in Bhandara, *Agricultural Marketing*, Vol. VIII, No. 2, July, 1965.

Choudhary, S., Stabilization of Agricultural Prices in India in the Context of Rapid Economic Development, *Indian Journal of Agricultural Economics*, Vol. XII, No. 1, January-March, 1957.

Chauhan, K.K.S. and Verma, R.C., Price Structure of Bajra in Jaipur, *Agricultural Marketing*, Vol. XII, No. 3, October, 1969.

Dandekar, V.M., Minimum Support Price for Foodgrain : Guidelines for a Policy and a Programme, *Artha Vijanan*, Vol. VII, No. 1, March, 1965.

Dandekar, V.M., Agricultural Price Policy : A Critique of Dantwada, *Economic and Political Weekly*, Vol. III, March, 1968.

Gadgil, D.R., Agricultural Price Policy - A State for Achieving Targets, *Yojana*, Republic Number, 1961.

Gangawar, A.C., Shrinivas and Rai, K.N., Marketing Pattern of Gram in Selected Markets of Haryana, *Agricultural Marketing,* Vol. XXVI, No. 1, April 1983.

Garg, J.S., Prasad, V. and Singh, G.N., An Approach to Determine Sugarcane Price, *Agricultural Marketing,* Vol. XV, No. 1, April, 1972.

George, M.V. and Ranjit Singh, Structure of Paddy Prices in Punjab, *Agricultural Marketing,* Vol. Nil, October, 1968.

Gill, K.S. and Johl, S.S., Gram Price Structure in Punjab, *Agricultural Situation in India,* Vol. XXV, No. 7, October, 1970.

Hosamani, S.B., Hugar, L.B. and Hiremath, K.C., Structure and Movement of Market Arrival and Price Behaviour of Cotton in Belgaum District, *Agricultural Marketing,* Vol. XXVIII, No. 2, October-November, 1985.

Indrasena Reddy, Price Behaviour of Agricultural Commodities, *Indian Journal of Marketing,* Vol. 23, No. 6-7, June-July, 1995.

Jairth, M.S. and Kansal, A.K., A Case Study of Price Behaviour of Himachal and Kashmir Apples in Delhi Market, *Agricultural Marketing,* Vol. XXIV, No. 2, July, 1981.

Johl, S.S., George, M.V. and Singh, A.J., Agricultural Prices in Punjab - A Policy Analysis, *Indian Journal of Agricultural Economics,* Vol. XXV, No. 1, January-March, 1970.

John W. Mellor, Functions of Agricultural Prices in Economic Development, *Indian Journal of Agricultural Economics,* Vol. XXIII, No. 1, 1968.

Kahlon, A.s., Stabilization of Agricultural Prices in India in the Context of Rapid Economic Development, *Indian Journal of Agricultural Economics,* Vol. XII, No. 1, January-March, 1952.

Kahlon, A.S. and Dwivedi, H.N., Seasonal Price Fluctuations in Agricultural Produce, *Agricultural Marketing,* Vol. VII, No. 1, April, 1965.

Kainth, G.S. and Balbir Singh, Pricing Efficiency in Foodgrains Marketing System-The Nature of Spatial and Temporal Price Relationship, *Agricultural Marketing,* Vol. XXIII, No. 1, April, 1980.

Kamala Devi, P., Price of Jowar in Southern zone Between 1950 and 1960. *Agricultural Situation in India,* Vol. XVI, No. 2, February, 1962.

Katre, R.S., Thakare, R.P. and Borade, S.G., Trend and Variability in the Price and Arrivals of Rice (Paddy) in Selected Markets of Maharashtra State, *Agricultural Situation in India,* Vol. XLVIII, No. 2, May, 1993.

Mcneely, John G., Some Problems of Price Analysis for Rice, *Artha Vijanan,* Vol. I, No. 2, June, 1959.

Mohd Muzammil, Trading Pattern and Price Behaviour, *Indian Journal of Marketing,* Vol. XIII, No. 9-10, March-June, 1983.

Mujumdar, N.A., Some Notes on the Price Policy Implications of State Trading In Foodgrains, *Indian Journal of Agricultural Economics,* Vol. XX, No. 1, January-March, 1965.

Naidu, M.R. and Goverdhan Reddy, A Critical Review of Jaggery Marketing at Anakapalli Regulated Market (A.P.), *Agricultural Marketing,* Vol. XXIV, No. 3, October, 1981.

Natarajan, B., Agricultural Price Fixation in Madras State, *Indian Journal of Agricultural Economics,* Vol. VI, No. 1, March, 1951.

Parthsarthy, H. and Mudahar, M.S., Foodgrain Prices and Economic Growth, *Indian Journal of Agricultural Economics,* Vol. XXXI, No. 2, April, 1976.

Pathak, H.N., Stabilization of Agricultural Prices, *Indian Journal of Agricultural Economics,* Vol. VI, No. 1, March, 1951.

Pushpavalli, A., Market Arrivals and Prices Behaviour of Groundnut in Tiruvannamalai Regulated Market (Tamil Nadu), *Indian Journal of Agricultural Economics,* Vol. XXXVIII, No. 3, July-September, 1993.

Rao, V.M., Market Arrivals and Price Movements, *Kurukshetra,* Vol. XLI, No. 6, March, 1993.

Schultz, T.W., Economic Effect of Agricultural Programme, *The American Economic Review,* Vol. XXX, No. 5, February, 1941.

Sharma, R.K. and Nandal, D.S., Structure of Arrivals and Prices of Gram in Hissar District of Haryana, *Agricultural Marketing,* Vol. XVII, No. 2, July, 1974.

Sinha, Surendra Prasad and Verma, B.N., A study of Trend and Variations in the Prices of Foodgrains in Bihar with Special Reference to Prices of Cereals between 1956-1968, *Indian Journal of Agricultural Economics,* Vol. XXVI, No. 4, October-December, 1971.

Venkataramanan, L.S. and Muralidharan, M.A., Seasonal Price Movements, Market Arrivals and Returns to Storage in Wheat Markets, *Indian Journal of Agricultural Economics,* Vol. XXVII, No. 2, April-June, 1972.

Vishwanath and Singh, R.P., Agricultural Prices in India, *Agricultural Marketing,* Vol. XV, No. 4, January, 1973.